T0361890

ROUTLEDGE LIBRARY EDITIONS:
AGRICULTURE

Volume 2

AGRICULTURE AND THE DEVELOPMENT PROCESS

AGRICULTURE AND THE DEVELOPMENT PROCESS

A Study of Punjab

D. P. CHAUDHRI AND
AJIT K. DASGUPTA

Routledge
Taylor & Francis Group

LONDON AND NEW YORK

First published in 1985 by Croom Helm Ltd

This edition first published in 2020
by Routledge
2 Park Square, Milton Park, Abingdon, Oxon OX14 4RN

and by Routledge
52 Vanderbilt Avenue, New York, NY 10017

Routledge is an imprint of the Taylor & Francis Group, an informa business

British Library Cataloguing in Publication Data
A catalogue record for this book is available from the British Library

ISBN: 978-0-367-24917-5 (Set)
ISBN: 978-0-429-32954-8 (Set) (ebk)
ISBN: 978-0-367-25035-5 (Volume 2) (hbk)
ISBN: 978-0-429-28575-2 (Volume 2) (ebk)

Publisher's Note
The publisher has gone to great lengths to ensure the quality of this reprint but points out that some imperfections in the original copies may be apparent.

Disclaimer
The publisher has made every effort to trace copyright holders and would welcome correspondence from those they have been unable to trace.

Agriculture and the Development Process

A STUDY OF PUNJAB

D.P. Chaudhri and Ajit K. Dasgupta

CROOM HELM
London • Sydney • Dover, New Hampshire

©1985 D.P. Chaudhri and Ajit K. Dasgupta
Croom Helm Ltd, Provident House, Burrell Row,
Beckenham, Kent BR3 1AT
Croom Helm Australia Pty Ltd, First Floor,
139 King Street, Sydney, NSW 2001, Australia

British Library Cataloguing in Publication Data
Chaudhri, D.P.
    Agriculture and the development process: a
    study of Punjab.
    1. Punjab (India)—Economic conditions
    2. Agriculture—Economic aspects—India—
    Punjab
    I. Title   II. Dasgupta, Ajit K.
    330.954'55204   HC435.2.Z7P8
    ISBN 0-7099-3408-4

Croom Helm, 51 Washington Street, Dover,
New Hampshire 03820, USA

Library of Congress Cataloging in Publication Data
Chaudhri, D.P.
    Agriculture and the development process.

    Bibliography: p.
    Includes index.
    1. Agriculture – Economic aspects – India – Punjab –
History. I. Dasgupta, Ajit Kumar. II. Title.
HD2075.P8C5 1985   338.1'0954'552   84-27507
ISBN 0-7099-3408-4

Printed and bound in Great Britain by
Biddles Ltd, Guildford and King's Lynn

CONTENTS

CONTENTS

LIST OF TABLES AND FIGURES

LIST OF TABLES AND FIGURES

LIST OF TABLES AND FIGURES

LIST OF TABLES AND FIGURES

PHYSIOGRAPHY OF PUNJAB

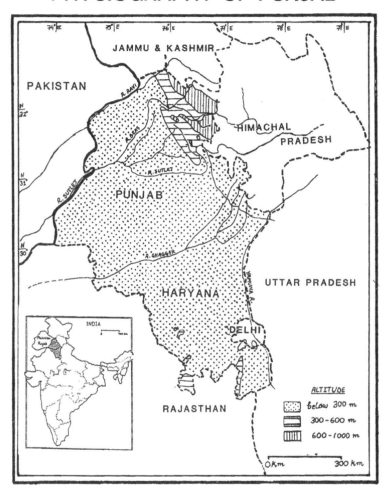

For Prem and Shipra

PREFACE

Much has been written on 'the Green Revolution' in
Punjab. Some have praised its achievements in
speeding up growth. Others have blamed it for
accentuating inequality. In this study, we have
looked at agricultural growth in Punjab in a
historical perspective. We have also tried to see
different aspects of economic development in the
region in an integrated way. By doing so we hoped
to learn something about both what agricultural
growth can contribute to the development process and
what it cannot.

Punjab was divided when India was partitioned
in 1947. The boundaries of Indian Punjab were
further adjusted in 1953 and it was subdivided into
Punjab and Haryana states in 1966. Our data series
are so adjusted as to apply to the geographical
region of Indian Punjab as of 1960-61. In some
cases such adjustment was not possible because of
the lack of data and for these, our region
represents the present Punjab and Haryana states
combined. However, the difference does not
materially affect any of the conclusions reached.
On occasions we have used 'Punjab' to refer to some
other entity, for example in ancient India, but this
should be clear from the context.

We have drawn mostly on published data from
standard sources, such as the National Sample
Survey, the Census, the Farm Management Survey and
National Income Statistics, and also used some
unpublished material. Terminals for our data series
differ, because of problems of availability. The
more important limitations of the data are indicated
in the text itself or in appendices. On some
questions we have also relied on our own
observations made during visits to different areas
of Punjab over the past two decades and, in case of
one author, during extended field-work carried out
in the late 1950s and the late 1960s in a number of

villages.

Our immense debt to scholars who have written on Punjab, agriculture or development should be clear from the text. We gratefully acknowledge the facilities for research and the stimulating academic environment of the Institute of Economic Growth, Delhi, and the Development Studies Centre of the Australian National University, Canberra, which made this work possible.

We thank Max Corden for suggesting this enterprise; G.S. Bhalla, Ramesh Bhatia, Ashish Bose, Kanchan Chopra, Shipra Dasgupta, K. Krishnamurti, Satya Pal and R.M. Sundrum for discussion of specific points; Anita Kumari for research assistance; and N.A. Kazmi for computations. H.L. Mehta, M. Boock, Joyce Barrett and Anne Cappello prepared several versions of the typescript with patience and precision; Chris Blunt and Peter Oates assisted in preparation of the final typescript at the ANU computer. All errors are our own.

Chapter I

INTRODUCTION

*Do farmers prosper? Are reservoirs in the State large, full
and well-distributed so that crops do not suffer by a lack of rain?*

Mahābhārata, 2.5. 66-67

This chapter has four sections. The geography of
Punjab and its early history are briefly described
in the first section and some core aspects of its
development in the second section. Pre-independence
agricultural growth in Punjab is discussed in the
third section while the last describes the impact of
partition.

1 Geographical and Historical Setting

Punjab literally means 'five waters': those of the
rivers Sutlej, Beas, Ravi, Chenab and Jhelum,[1] the
main tributaries of the Indus. They arise in the
Himalayan ranges to the north, are fed by melting
snows and monsoon rains, and flow across a vast
alluvial plain. The eastern part of Punjab, which
became Indian Punjab on the partition of India in
1947, is strictly speaking the land of the three
rivers (Sutlej, Beas and Ravi) for by far the
greater part of it is covered by alluvial and
wind-blown deposits of the plains of these rivers.
    Its subtropical position of 70' to 78'
longitude and 31' to 33' latitude provides for four
distinct seasons and only about 50 cms of annual
rainfall with a range of 20 to 70 cms in different
regions. Rainfall is determined by the cycle of
monsoons which in turn affects river flows. Water
is the major resource and also a major constraint.
Humidity and rainfall levels are more in line with
the semi-desert lands of the Middle East and without
major water management and artificial irrigation

1

systems could only allow a short crop calendar and
limited agricultural development. Because of this
irrigation has always been a central concern of the
state.[2]
       Historically, the Punjab region was among the
first to develop a civilisation based on settled
agriculture, viz. the Harappan or Indus Valley
civilisation of the third millennium BC,
acknowledgedly 'one of the great historic
achievements of the human race'.[3] It was an urban,
literate civilisation with a developed agricultural
base providing surplus food which was stored in huge
granaries in the metropolitan centres in large
enough quantities to maintain the urban population
in bad years as well as good. This was made
possible by an organised system of irrigation,
largely by inundation, and the use of a light plough
or harrow.[4] Wheat, barley and cotton were the
principal crops and leguminous plants and field peas
were cultivated as well. Domestic animals included
sheep, goats, cattle and fowls,[5] and possibly also
pigs and buffaloes. Subsequently, with the decline
and fall of Harappan civilisation and the ascendancy
of Indo-European speaking, copper and bronze using,
pastoral-nomadic immigrant tribes, sometimes called
'Aryans', Punjab still retained a pre-eminent
position. The rivers of the Rig-Veda are those of
the Punjab, and for this reason it has been
described 'as the centre of Rig-vedic geography'.[6]
Later, with the development of a technology based on
the use of iron implements and the cultivation of
rice, the centre of North Indian civilisation moved
inevitably to the lower Ganges Valley. Punjab
continued to be an important region but no new
thrust towards an urban civilisation emerged. It
remained almost entirely an agrarian society.

2 Core Aspects of Development

The essential feature of the Punjab economy remained
unchanged for a long historical epoch.[7] The basic
instrument of production remained the light wooden
plough, sometimes iron-tipped and sometimes not.
The scarcity of rainfall made cultivation heavily
dependent on irrigation, which was done by water
drawn from tanks, wells and river channels. Lift

irrigation was operated both manually and by
bullocks; and from early times the bullock-drawn
Persian wheel was used for this purpose by the more
affluent land owners. Crop rotation closely
followed the soil and climatic pattern prevailing in
each particular region. The system, primitive as it
might appear, was well suited to local conditions
and was by no means inefficient by standards of
traditional agriculture, the seed yield ratio of
wheat during the mediaeval period was probably more
favourable than that in mediaeval European
agriculture. The system even allowed for expansion
of output which indeed did occur in certain periods.
This was achieved by an extension of acreage,
usually brought about by more irrigation. At some
times, and in certain areas, farmers changed their
crop mix in favour of higher value crops, for
example from jowar and bajra to wheat and cotton,
probably in response to changes in relative prices
and differential crop-specific levies of land
revenue imposed by the Mughal state. What the
system lacked however was any incentive for the
adoption of innovations which could raise the yield
rate or productivity of labour. This was reflected
in the long-run stability of yield rates.
 As in other systems of traditional agriculture
the basic unit of production was the family farm.
The system was neither entirely closed nor entirely
self-sufficient as often supposed. Theories of 'the
Indian village community' which assert that land was
necessarily held in common and 'owned' by the state
are also not supported by results of recent
historical research. On the contrary there is much
evidence of a complex hierarchy of rights to the use
of land and even of considerable differences in the
size of holdings in the same village. Production
for the market especially in cotton and sugarcane
was known from early times. ` There was also some
limited development of trade and commerce in grains;
and fairs held in bigger villages or nearby towns on
the occasion of religious festivals served as
channels of trade. Moneylending developed from a
fairly early period and both moneylender and
merchant had an important social position in the
village.
 The economic surplus generated in agriculture
was extracted by the state in the form of land

revenue; and only a little of it was reinvested in rural public works. Its function was to finance military expenditure and courtly splendour. While the system remained stable for a very long period, by late Mughal times it had already come under heavy strain, showed <u>inter alia</u> by a flight from the land.[8]

The short period of Sikh rule saw a simplification of the land revenue system, a drastic curtailment in the powers of the Zamindar (landlord), and a reduction in the number of intermediaries between the direct producer and the state.[9] This developed further under British rule when in effect a <u>ryotwari</u> system of peasant proprietorship became established in many parts of Punjab. The cultivation of cash crops increased especially in West Punjab and exports, especially of cotton, both to foreign countries and to other parts of the country became more important. There was considerable increase in tenancy and moneylending; and the village economy became much more monetised than before. Except in a few isolated pockets there was not much impact on productivity, however, despite the construction of a fairly extensive system of canal irrigation.

Two aspects of agricultural development in Punjab deserve emphasis. One is the importance of irrigation, to which we have already referred. Historically, the development of irrigation has been a slow and continuously evolving process from Harappan times to the present. Seasonal flood inundation by canals and the Persian wheel had long provided traditional instruments of irrigation in this region. These were improved during Mughal rule by the introduction of weirs, barrages and canals which helped to control river flows. They were considerably improved by the British who also built storage dams and developed canal colonies, though these were almost entirely in West Punjab. The construction of the Bhakra Nangal system and later of very large numbers of private and public tubewells are the major achievements since independence.[10]

The other aspect is the nature of the links between the economy and society of Punjab and the outside world. One such link, from pre-historic times, was trade [11];but this can hardly be called

4

Introduction

a distinctive feature of the Punjab, for the impact
of external trade in such regions as Saurashtra and
Malabar was surely more pronounced.
War provided a stronger link. Invaders of
India, from Darius to Babar, came mostly through the
passes to the west and north-west of India, and went
on, with distressing regularity, to cross the Punjab
plains. The unsettled quality of life which this
implied could have helped to make people more mobile
and receptive to outside ideas and influence, less
hierarchical and caste-bound, more adaptable and
responsive to challenge.[12]

3 Punjab Agriculture 1906-1946

Agricultural growth in pre-independence Punjab has
frequently been regarded as the one indisputable
success story in the bleak record of Indian
agriculture under the British Raj. The evidence
hardly supports such an optimistic view.
The classic study of agricultural growth in
pre-independence India is Blyn's (1966). He was,
however, concerned more with the all-India pattern
than with regional growth as such; and more with
deriving reliable estimates of the time series of
output, acreage and yield than with a rigorous
analysis of the series themselves.
As far as Punjab is concerned, he found a
positive growth rate in output and acreage and to a
lesser extent in the yield rate, for each of the
categories all crops, food crops and non-food crops
during the period 1891-92 to 1945-46 as a whole,
although growth rates, especially of food, tended to
decline over time.
The significance of this growth, according to
Blyn, is essentially that it helped to maintain the
all-India output of food at a stable level. After
1920, it was the positive growth rates of foodgrains
output in the Punjab and Bombay that balanced the
decline in the rest of India, the rate of growth at
the all-India level being close to zero.
Two other studies are relevant, those by
Rajkrishna (1964) and Chander Prabha (1969).
Rajkrishna presents estimates of the rate of growth
of aggregate output of eleven major crops, viz.
cotton (American), cotton (Desi), maize, sugarcane,

5

bajra, jowar, rice, wheat, barley, rape and mustard
and gram for the 29 districts of undivided Punjab
during 1913-14 to 1945-46. He also tries to
estimate how far the growth of agricultural output
in Punjab during this period could be explained by
increases in factor supplies or in productivity:
his main result is that it was essentially due to
the extension of irrigated area, yield rates
remaining about the same.

Chander Prabha estimates growth rates of output
(the same crops being covered as in Rajkrishna's
study) in each district of the Punjab except Simla
during 1906-07 to 1945-46 and, by observing their
geographical pattern, concludes that the growth rate
in the region which later became Punjab (Pakistan)
must have been significantly higher than in that
corresponding to Punjab (India), though she does not
actually estimate the two growth rates concerned.
This is done by Dasgupta (1982) who has analysed
crop output and acreage separately for West and East
Punjab and derives the corresponding growth rates.
This was done by fitting semilog regressions to time
series of total acreage, and of the value of total
output at Punjab farm harvest prices (average
1924-25, 1926-27) of ten major crops, viz. rice,
wheat, barley, jowar, bajra, maize, gram, sugarcane,
American cotton and Desi cotton. The analysis is
for the period 1906-07 to 1941-42; the use of an
earlier year as the starting point involves serious
problems of comparability while choosing 1941-42 as
the terminal year helps to exclude abnormal wartime
influences, that of the Grow More Food campaign in
particular.[13] The main result is presented in
Table 1.1.

Introduction

Table 1.1 COMPOUND GROWTH RATES, OUTPUT AND ACREAGE
         IN WEST AND EAST PUNJAB, 1906-07 TO
         1941-42
         (Per cent per year, ten major crops)

	Growth rate of output	Growth rate of acreage
West Punjab	1.2	0.8
East Punjab	0	0

Note: The basic data and the method of estimation
      of growth rates are discussed in Dasgupta
      (1982).

    Clearly, agricultural growth in Punjab was
confined to its western part. East Punjab remained
stagnant. Since West Punjab had three-quarters of
the total canal-irrigated area of Punjab including
all the great canal colonies, this is not perhaps a
particularly surprising result.
    Table 1.1 does not tell us much about the trend
in yield rates. Some idea of this can be had from
the figures presented in Table 1.2, which suggests a
decline in the yield rates of major crops in East
Punjab, except in the case of wheat and Desi cotton
(for which there was a rise) and sugarcane (the
yield rate of which remained the same).

    Further, the cropping pattern in East Punjab
was less favourable than in West Punjab, in the
sense that the proportion of irrigated area under
high-valued crops was lower.[14]
    While agriculture stagnated, there was little
urbanisation and manufacturing activity probably
declined, leading to an increase in the proportion
of the workforce in agriculture while that in
manufacturing fell sharply (see Table 1.3).

4 Partition and Punjab on the Eve of Independence

Perhaps the most traumatic shock that Punjab has

7

Table 1.2 CROP YIELD RATES IN WEST AND EAST PUNJAB
(Kilogrammes per hectare)

| | West Punjab | | East Punjab | |
	Average over 5-year period ending 1910–11	Average over 5-year period ending 1941–42	Average over 5-year period ending 1910–11	Average over 5-year period ending 1941–42
Rice	185	218	173	169
Wheat	140	144	136	160
Barley	136	127	123	115
Jowar	70	62	62	29
Bajra	78	70	62	25
Maize	152	173	156	144
Gram	103	74	115	82
Sugarcane	263	317	329	329
Cotton (Desi)	82	152	140	173

Source: Dasgupta (1982)

Table 1.3 OCCUPATIONAL DISTRIBUTION OF MALE WORKING
FORCE IN INDIAN PUNJAB 1911 AND 1951
(Per cent of Total)

	1911	1951
Agriculture and allied activities	63.1	67.6
Mining and quarrying	0.1	–
Manufacturing	14.4	7.9
Construction	1.5	1.0
Trade and commerce	6.2	8.6
Transport, storage and communication	2.2	2.2
Other services	12.5	12.7
Total	100.0	100.0

Source: Census of India, 1961, Paper 1, Delhi
(1967).

8

experienced in its long history was its partition on religious lines on the eve of independence in 1947. The communal carnage that accompanied partition affected most families. Three million Hindu and Sikh refugees poured into Indian Punjab within a single year and about an equal number of Muslims were uprooted. This was one of the largest such migrations in human history.

To the people of both Punjabs, the partition was a traumatic experience, but it was also a challenge. The problem was not just one of resettling the refugees and creating opportunities for their productive employment, though this was formidable enough. For Indian Punjab, there was an even more fundamental question: how to initiate a process of growth in an agriculture which was rain-fed, traditional and had long been stagnant? The response to it provided the impetus to agricultural growth in Punjab for the next quarter of a century. The response was both political and economic and was made, in the first instance, by the state and society of India, but equally it was an individual response by inhabitants of this region. The Bhakra Dam was built with Indian resources, by the skills of her engineers and the hard labour of her construction workers.[15] Canal irrigation based on Bhakra water was supplemented by rural roads and electricity, land consolidation and village schools, regulated markets and agricultural research. Farmers, especially medium and large farmers, responded by growing more crops on the same land, by changing traditional cropping patterns in favour of more profitable crops and crop rotation, by using chemical fertilisers and improved farm practices, by adopting new high-yielding varieties of crops when they became available, by investing in tubewells, tractors and other productive equipment. The result was an impressive growth in agricultural output which finally broke the age-old spell of agricultural stagnation. The nature and magnitude of this growth will be discussed in our next chapter and the building of the infrastructure in Chapter III.

Introduction

1. The Sanskrit names are Sutudri, Vipasa, Parushni, Asikni, and Vitasta, respectively. The corresponding name for Indus is Sindhu.
2. Cf. 'nearly all the dynasties which have ruled over the Five Rivers have done something for irrigation; nearly every district possesses flowing canals or else the ruins of ancient water courses',General Report on the Administration of the Punjab for the years 1849-50 and 1850-5 , London, printed for the Board of Directors of the East India Company, 1854, p.96, quoted in Michel (1967). It may be noted that Punjab was annexed by the British in 1848.
3. Clark (1969,p.209).
4. Kosambi (1965), Thapar (1975).
5. According to Basham (1979,p.26), 'Perhaps the most widely appreciated of prehistoric India's gifts to the world is the domestic fowl'. This is possibly a somewhat quixotic judgement, but the gift was that of the Punjab.
6. Allchin (1968,p.154). See also Kosambi (1965).
7. Historians are still far from agreeing on what the essential features were. The best general account is to be found in the Cambridge Economic History of India, volume 1, c.1200 - c.1750, edited by Tapan Raychaudhuri and Irfan Habib, Cambridge University Press,1982; and volume 2, c.1757 - c.1970, edited by Dharma Kumar, Cambridge University Press, 1983, which also provides a detailed bibliography.
8. See Raychanduri and Habib (1982), especially the contributions by Habib on the system of agricultural production in Mughal India (chapter VIII, section 1) and on agrarian relations and land revenue in North India (chapter IX, section 1). Mughal rule in India was established in 1556.
9. See Kessinger (1974) who describes this as constituting a fundamental change in the structure of rural society in Punjab. The history of Punjab during the second half of the eighteenth century was essentially that of the rise of the Sikhs as a military and political power in a struggle against Mughal authority; but the Sikh misls also fought among themselves for control. Rangit Singh assumed

the title of Maharajah of Punjab in 1799 and soon
established both his military ascendancy and a
united administration. Sikh rule ended with the
first Sikh war of 1845-46.
10. Their impact on agricultural growth in
Punjab will be discussed in Chapter II.
11. Trade with Sumer and the ports of the
Persian Gulf was especially important, leading one
scholar to remark: 'From the third millennium B.C.
the Punjab formed the most eastern region of the
ancient near East, connected by sea and land with
the west and by trade, transgressing the natural
north and western frontiers of the Indian
subcontinent' (Ruben ,1974). Other scholars would
assign a more independent role to the Indus Valley
civilisation (cf. Clark, op.cit.).
12. Just how important this was is of course
impossible to determine precisely. It could be
argued , for example , that the Jats of the Punjab
had been distinguished from early times by an
absence of social or economic stratification
(Habib,1974). On the other hand , the egalitarian
emphasis of Sikh teachings has itself been regarded
as a primary reason for their allegiance to Sikhism
(McLeod, 1976,pp.11-12). Again some scholars have
emphasised the persistence of caste-based rituals
and conventions among Sikhs, which would appear to
contradict our hypothesis that Sikhism was
egalitarian, but McLeod explains this, in our
judgement convincingly, by pointing to the
distinction between the 'vertical' and 'horizontal'
aspects of caste – the former being represented by
the varnas and the latter by jati and gotra. 'We
can affirm once again their (i.e. the Gurus)
apparent acceptance of the horizontal relationship,
an acceptance unmistakably demonstrated by their
willingness to observe customary marriage
conventions. What they were apparently concerned to
deny was the justice of privilege or deprivation
based upon notions of status and hierarchy. They
were, in other words, opposed to the discriminatory
aspects of the vertical relationship while
continuing to accept the socially beneficial pattern
of horizontal connections.' (McLeod, op. cit.pp.
90-91).
13. This campaign was started in 1942 with
objectives of encouraging a switch from commercial

cash crops to food crops, promoting more intensive cultivation of land and bringing currently non-cultivated land under cultivation. It had an appreciable but shortlived effect on food crops output and area in the Punjab. See Report of the Grow More Food Enquiry Committee, Government of India, 1952.

14. During the five-year period ending 1940-41 the average percentage shares of total irrigated area in East Punjab were as follows: wheat 22.66; maize 15.68; gram 14.93; bajra 13.47; cotton (Desi) 4.87; jowar 3.51; barley 3.43; rice 3.05; sugarcane 2.03; cotton (American) 0.16; fodder crops 15.50; others 10.71. The shares of wheat and of gram in East Punjab were approximately half and twice respectively of those in West Punjab.

15. See Michel (1967,pp. 380-81), who points out the differences in policies adopted by the Indian and Pakistan governments regarding dependence on foreign resources in building up their irrigation infrastructure, and its long-term consequences.

Chapter II

GROWTH OF AGRICULTURAL OUTPUT AND INPUTS

Oh sister, give me some canal land
Oh brother, sow some land on the canal

Punjabi folk song

Modern industry alone, and finally, supplies, in machinery,
the lasting basis of capitalistic agriculture...

Karl Marx

While the remarkable growth in agricultural output
which has occurred in post-independence Punjab,
especially since the mid-1960s, has attracted
considerable attention, there have been few attempts
to analyse the time series of output in a systematic
way. The objective of this chapter is to provide
elements of such an analysis. Just how fast did
agricultural output increase? Was the rate of
growth since the mid-1960s all that different from
that which occurred in the 1950s? Has the Green
Revolution fizzled out? Did agricultural growth
remain confined to wheat? These are some of the
questions that our analysis will attempt to answer.
    The chapter is organised in five sections. The
first provides estimates of the rate of growth of
agricultural output. Trends in land, labour and
capital are described successively in the next three
sections. Factor proportions and productivity
changes are discussed in the final section.

1 Growth Rates of Output 1950-1979

We have used four different series of output data
which, expressed in index number form, are given in
Table 2.1. (A) represents a 'value added' measure
of agricultural output, including livestock
products, in constant prices; (B), (C) and (D) on
the other hand represent real gross value of output
rather than value added and refer to crop production
only, i.e. exclude livestock activities. The
primary data underlying the indices and methods used

13

Growth of Agricultural Output and Inputs

Table 2.1 INDEX NUMBERS OF AGRICULTURE OUTPUT IN PUNJAB
1950-51 TO 1978-79

(1960-61 = 100)

Year	Output net of material costs (1970-71 prices)	All Crops (official index)	10 Major Crops 1959-60 - 1961-62 Prices	10 Major Crops 1976-77 - 1978-79 Prices
	(A)	(B)	(C)	(D)
1950-51	n.a.	57	55	53
51-52	n.a.	58	54	51
52-53	n.a.	62	69	67
53-54	n.a.	60	76	75
54-55	n.a.	78	82	83
55-56	n.a.	71	77	78
56-57	n.a.	82	87	88
57-58	n.a.	84	87	87
58-59	n.a.	93	99	104
59-60	n.a.	89	94	91
1960-61	100	100	100	100
61-62	n.a.	94	96	89
62-63	n.a.	91	90	88
63-64	n.a.	96	93	89
64-65	n.a.	116	114	103
65-66	107	104	98	90
66-67	n.a.	116	111	103
67-68	n.a.	154	146	138
68-69	n.a.	149	148	124
69-70	n.a.	186	186	170
1970-71	162	191	208	170
71-72	161	195	192	170
72-73	156	188	184	163
73-74	162	189	183	151
74-75	162	190	181	158
75-76	182	223	217	193
76-77	191	235	219	201
77-78	207	256	246	217
78-79	n.a.	286	274	240

Source: See Appendix B for source of data and methods of estimation.
n.a. denotes not available.

in deriving index numbers are discussed in Appendix
B.

Growth rates over this entire 1950-51 to
1978-79 period and in successive sub-periods, which
were derived from these indices in a number of
different ways, are presented in Tables 2.2(a) and
2.2(b). The reason for using a variety of data
series on output in estimating growth rates, rather
than choosing one only, is essentially that we wish
to explore how sensitive measured growth rates are
to the choice of an output index. Thus, for
example, the official index of crop production
(Series B) has a wide coverage in terms of crops
included but is based on relative prices of a fairly
early phase of agricultural development. Hence we
built up the series (C) and (D) which include the
major crops, representing about 85-90 per cent of
the value of all crops in a normal year, but two
different sets of price weights, one early (1959-60
to 1961-62) and another the latest for which data
are available (1976-77 to 1978-79). Series (A) on
the other hand gives a broken series but it is the
only one available on value added output (net of
material costs), and we have used it because a
comparison of the growth rates of value added with
that of gross crop production may help to bring out
the role of intermediate inputs in the process of
economic growth in the Punjab.

Some caveats are necessary, for growth rates
computed over short periods are notoriously
unreliable [1], and those of agricultural output,
which varies with weather, particularly so.[2] They
tend to be sensitive to the choice of the initial
and terminal years. Small errors in basic data can
have a large effect. Their magnitude depends on the
measure of growth rate used; and even the value of
a compound growth rate itself can vary with the
method used in its estimation. It follows that
small differences between different short-period
growth rates are quite devoid of significance.

Limitations of growth rates notwithstanding, a
fairly clear pattern of growth emerges from our
results. The decade of the 1950s saw rapid growth,
at a compound rate of over 5 per cent per year. A
period of stagnation followed, broken in turn by a
quinquennium beginning 1965-66, in which compound
yearly growth rates of around 11-12 per cent in

15

Growth of Agricultural Output and Inputs

Table 2.2(a) COMPOUND GROWTH RATES OF AGRICULTURAL OUTPUT
IN PUNJAB: 1950-51 TO 1977-78
(Per cent per year)

Period	(A) Output net of material cost	(B) All Crops, official index	(C) 10 Major Crops, 1959-60 – 1961-62 Prices	(D) 10 Major Crops, 1976-77 – 1978-79 Prices
1	2	3*	4*	5*
I 1951-52 - 1960-61	n.a.	5.3	5.7	5.6
II 1960-61 - 1965-66	1.3	2.2	2.2	1.3
III 1965-66 - 1970-71	8.8	12.7	12.5	11.4
IV 1970-71 - 1974-75	0	1.3	-0.1	-0.4
V 1974-75 - 1977-78	8.5	8.8	8.2	9.5
VI 1951-52 - 1977-78	4.4 **	5.9	5.6	5.3

Notes: * Figures represent yearly compound growth rates per cent
between three-year moving averages of the Index Numbers
shown in Table 2.1 centred on the initial and terminal
years of each period shown here.

** for the period 1960-61 - 1977-78.
Source: Table 2.1.

16

gross output and nearly 9 per cent in value added
were achieved. The early 1970s marked a lull with
near zero growth rates. In 1974-75 began another
period of rapid growth,the compound growth rates
over the next three years being about 8-9 per cent
per year. The differences in growth rates between
the periods mentioned remained substantial whichever
set of estimates are used and cannot be regarded
simply as a statistical illusion.

Tables 2.2(a) and 2.2(b) show that growth rates
based on the indices (B), (C) and (D) are fairly
close together most of the time, and it does not
matter much which one we use for our analysis. On
the other hand, a comparison of growth rates of
value-added derived from (A) with those of crop
production based on any of the other indices is
instructive. The value added series (A) is not
available for the 1950s but for both the other
periods of substantial growth, viz. III and V, it
shows growth rates that are significantly lower than
that of gross crop production. This reflects the
increasing use of purchased intermediate inputs such
as chemical fertilisers, pesticides and fuel which
in turn implies greater linkages of Punjab
agriculture with the rest of the economy, for such
inputs come not only from outside agriculture but
also, for the most part, from outside Punjab.[3]

We conclude this discussion with a comment on
our periodisation. Tables 2.2(a) and 2.2(b) provide
growth rates for five different sub-periods. Since
any period can be broken up in different ways, and
the growth rates will vary accordingly, an element
of arbitrariness is necessarily involved. Our
choice of periods is not merely capricious however.
First, they reflect distinct phases shown by the
data themselves. Secondly, they correspond to
certain stages of post-independence economic
development in the Punjab. The 1950s show the
extension of agriculture under the impact of the
Bhakra irrigation system. The mid-1960s mark the
Green Revolution in wheat and the period from
1974-75 of one in rice. The two other periods in
our scheme are essentially transitional.

So far, we have discussed the growth of
aggregate output in agriculture. We shall now
consider briefly changes in its composition. Some
idea of this may be had from Table 2.3 which shows,

Growth of Agricultural Output and Inputs

Table 2.2(b). SIMPLE GROWTH RATES OF AGRICULTURAL OUTPUT IN PUNJAB,
1950-51 TO 1978-79
(Per cent per year)

Period	(A) Based on index of output net of material cost	(B) Based on All Crops official Index	(C) Based on index of 10 major crops 1959-60 - 1961-62 prices	(D) Based on index of 10 major crops, 1976-77 - 1978-79 prices
I 1951-52 - 1960-61	-	6.1	7.1	7.3
II 1960-61 - 1965-66	1.4*	1.5	0.03	-1.6
III 1965-66 - 1970-71	10.28*	13.64	16.7	15.1
IV 1970-71 - 1974-75	0.00*	-0.16	-2.7	-1.7
V 1974-75 - 1978-79	9.26*+	10.85	11.1	11.2
VI 1950-51 - 1978-79	6.29*++	6.40	6.72	6.37

Notes: The figures represent yearly percentage growth rates and
         were derived by taking the arithmetic mean of
         percentage changes in the Index Numbers shown in table
         2.1 in successive years within each period, except for
         those marked with an asterisk in which case the
         percentage growth over the period was divided by number
         of years involved.
       + This is for the period 1974-75 - 1977-78.
      ++ for 1960-61 - 1977-78.
Sources: as in Table 2.1.

18

for each of the nine major crops of this region, its
changing percentage share in the constant-price
value of their aggregate output since 1950-51. The
shares of rice and of wheat have both increased
considerably in this period. Together they
accounted for about 40 per cent of the total in
1950-51. In 1978-79 the corresponding figure was
nearly 75 per cent. The increase in the share of
wheat, which has always been a major crop in this
region, occurred mostly during the second half of
the 1960s, which confirms the wide-spread general
impression that the Green Revolution was in the
first instance a wheat revolution. Subsequently,
however, the share of wheat has remained stable
while rice has made rapid gains. For the period as
a whole the trend is clearly towards greater
specialisation, and in favour of the rice-wheat
rotation in particular. The share of gram has also
declined sharply, but this had started earlier, i.e.
in the mid-1950s.

These changes in the composition of total crop
output were due to different rates of growth of
different crops. Here, there is an important
difference in the nature of agricultural growth
occuring during the decade of the 1950s and that
since the mid-1960s. In the former period most of
the major crops had high rates of growth. The
compound growth rates per cent per year, computed
from three-year moving averages of crop output in
physical units over 1951-52 to 1960-61, were 8.07
for gram , 6.76 for sugarcane, 6.40 for rice, 5.01
for wheat and 4.95 for maize. This is not so for
the period since the mid-1960s. Growth has occurred
mostly in rice and wheat. For other crops growth
rates have been much less marked and in some cases
even negative. This was indeed the mechanism by
which greater specialisation was achieved.

To sum up, the growth of agricultural output in
post-independence Punjab has been both sustained and
substantial. The cropping pattern has shifted
towards more high-valued crops. The long stagnation
that marked the colonial period has been decisively
broken. Growth rates have been about two to three
times as high as those in India as a whole and
compare favourably with those experienced in any
region of comparable size and population over a
period of similar duration. In the concluding

Table 2.3   PERCENTAGE SHARE OF PRINCIPAL CROPS IN PUNJAB 1950-51 TO 1978-79
(Per cent of value in 1969-70 to 1971-72 farm harvest prices)

Year	Rice	Wheat	Barley	Jowar	Maize	Gram	Bajra	Sugarcane	Cotton	Total
1950-51	3.84	36.62	2.47	1.76	3.96	26.49	8.18	13.42	3.26	100.00
1955-56	3.20	33.42	2.40	0.52	5.02	34.38	8.38	11.26	4.42	100.00
1960-61	5.00	35.20	1.33	0.62	5.64	29.07	2.99	15.54	4.61	100.00
1965-66	5.90	40.53	2.21	0.42	7.57	12.16	3.18	21.60	6.47	100.00
1970-71	7.07	56.70	0.93	0.38	5.22	9.97	6.14	10.73	3.86	100.00
1975-76	11.16	54.32	1.67	0.21	4.67	9.37	3.99	9.87	4.74	100.00
1978-79	18.32	56.40	0.71	0.12	2.74	7.63	1.85	7.78	4.45	100.00

Note: The American and Desi varieties of cotton were valued at their respective price and are
presented together.
Source: Statistical Abstract of India, various issues.

chapter, we shall discuss in some detail the agricultural growth of Punjab as compared with that occurring in Japan in the past and in other states in India more recently. We shall now turn to the question of how the change was brought about. We start by considering the contribution made by the primary factors of production: land, labour and capital.

2 Land Use and Cropping Intensity

The measurement of land as a factor of production involves a number of problems. Some of them are similar to those which arise in measuring capital while others are unique to land. Aspects of farming systems relating to location, natural fertility and ecology uniquely affect land even when it is measured in terms of physical area. Improvements of land brought about by investment are difficult to measure except in terms of costs of improvement which in turn involves complex problems of valuation.

As a measure of land we shall use net sown area, the area sown more than once in the same year being counted only once. A 'stock' concept of land is thus implied. This ignores land improvements due for example to bunding, consolidation and the extension of irrigation facilities, and increase in multi-cropping (sowing two or more crops in a year on the same land) made possible, inter alia, by irrigation is specifically excluded. Analytically, land improvements are properly conceived either as capital or as technical change embodied in such capital, and increases in cropping intensity (conventionally measured by the ratio of total cropped area to net sown area) as representing additional production possibilities associated with such change, rather than increases in the quantity of the factor 'land'. They will be discussed as such below. Increases in cropping intensity are shown in Table A.1.

The classification of land use has undergone changes in different states of India since independence. The system was standardised during the early 1960s.[4] On a comparable basis, five such components are distinguished in the land use statistics of Punjab, namely (i) forests, (ii) area

not available for cultivation, (iii) other
uncultivated land, (iv) fallow, and (v) net sown
area. Land use in this region during 1950-51 to
1978-79 according to these five categories is
reported in Table 2.4. This shows that net sown
area as a proportion of the total area has increased
from 69 to 83 per cent of the total geographical
area during this period. Part of the year to year
variation in net sown area may be attributed to the
timeliness or otherwise of the monsoon rainfall, but
an overall increase can only be the result of shifts
from some other categories of land use.

The area under forests has increased although
it continues to be a small proportion of the total.
The area not available for cultivation (called
ghairmumkin, i.e. uncultivable) is a broad category
which consists of barren and uncultivable land such
as deserts, rocks and mountains, sandy patches, land
affected by sub-soil moisture, unculturable alkaline
land and land in non-agricultural use, for example,
that occupied by buildings, roads and railways or
under water (rivers and canals). The percentage
share of this category has remained stable. It is
the other two items, namely (iii) other cultivated
land and (iv) fallow, which have declined, enabling
net sown area to expand. Apparently this reflects
both the cultivation of land which had gone out of
cultivation for five or more years, defined as
cultivable wastes, a major component of (iii), and
the running down of current fallow, of which item
(iv) almost entirely consists. With the spread of
irrigation and the increasing use of chemical
fertilisers the need to maintain the current fallow
for the purpose of maintaining soil fertility has
almost disappeared. Further decline of other
uncultivated or fallow land from their present
(1978-79) levels of 1.4 and 1.8 per cent
respectively is highly unlikely; and even if it
occurs the land would be taken up for
non-agricultural use. Thus net sown area appears to
have reached its peak in this region and future
changes can only be in the form of yearly
fluctuations around its present level.

Table 2.4 LAND USE IN PUNJAB 1950-51 to 1978-79
(Per cent of total)

Year	Forests	Land not available for cultivation	Other unculti-vated land exluding fallow	Fallow	Net Sown area	Total
1950-51	0.55	10.09	11.71	8.94	68.71	100
1955-56	0.67	11.26	9.36	5.97	72.74	100
1960-61	1.05	11.82	5.84	5.29	76.00	100
1965-66	1.83	12.04	3.46	6.91	75.76	100
1970-71	2.35	11.81	2.01	3.06	80.76	100
1974-75	3.37	10.68	1.68	3.61	80.65	100
1975-76	3.28	10.53	1.53	3.00	81.66	100
1978-79	3.48	9.97	1.41	1.79	83.35	100

Source: Statistical Abstract of India Various issues.

Growth of Agricultural Output and Inputs

3 Labour Supply and Labour Use

We shall measure labour by the number of male
workers reported to be in agriculture. Data are
from the Census and have been adjusted for
comparability. Our choice of measure for labour may
appear to be unduly restrictive but the data permit
no other.

For each individual the Census asks two
important questions: (1) does he participate in
economic activity, and (2) if so, what is its
nature? The answers to the first question give us
an estimate of the workforce, those to the second
its occupational distribution. The two are related,
for variation in definition of economic activity
affects the classification by occupational category;
for example, a narrow definition of economic
activity which excludes unpaid family workers would
lead to a serious underestimation of the workforce
in agriculture while it probably would not affect
that in industry.

Both the definitions, namely what constitutes
an economic activity and how an occupation category
is demarcated, have elements of arbitrariness.
There are, however, special difficulties in Indian
Census data, arising essentially from the failure of
different Censuses to adhere to a common set of
concepts and procedures. The 1951 Census used the
concept of gainful occupation as the basis of
economic activity. Receipt of income, however
small, provided it was regular, was the criterion
for being deemed economically active. Where two or
more members of a family household jointly
cultivated land and secured income thereby, each of
them, according to the 1951 Census instructions to
enumerators, was regarded as earning a part of that
income. Two other aspects of this approach are also
relevant. The economically active population was
enumerated by asking each individual to state his
occupation, and tabulations were made by selecting
those whose occupations as stated came within the
category of gainful work. In this sense it was the
person himself who stated his occupation and this
was done according to the 'usual status' concept
which could be, to some extent, independent of his
current activity.

In 1961 the concept of economic activity was

24

derived from the 'labour' approach, i.e. by the
criterion of regular participation in productive
work. In the case of agricultural activities this
meant work for more than one hour a day for the
major part of the working season. The criterion of
economic activity was work and persons obtaining
income without work were excluded. Subsequent
Censuses (1971 and 1981) have roughly followed the
same approach as used in 1961, although there are
some variations arising (especially in the Census of
1981) in the treatment of subsidiary occupations.

Two further difficulties add to the problem of
comparability between Censuses in respect of our
workforce data. There are a number of categories
which are unspecified or incompletely specified in
both the 1951 and 1961 Censuses. The unspecified
categories formed 3 per cent of the workforce in
1951 and 4 per cent in 1961. From the 1971 Census
this category is virtually eliminated and all
workers are assigned to specified occupations. The
actual occupations included under the unspecified
categories may have differed between the Censuses,
and even for 1971 those previously regarded as
unspecified may simply have been classified as
agricultural labourers, leading to an overestimate
of the inter-censal increase in their number.

Finally, the overlap between caste and
occupation is common to all Censuses in India. If
the actual occupation of an individual was different
from the traditional family caste occupation, it is
the latter which is more likely to have been
reported in the Census, especially if it carried a
higher status compared to the individual's actual
occupation. Since the influence of caste itself may
have changed in its strength, the errors on this
account need not necessarily be uniform. However,
since in the Punjab the influence of caste has
weakened over a very long period, this particular
difficulty may not be as acute as in some other
parts of India.

These difficulties in Census data imply that
one has to treat them with caution. However, it is
generally agreed by scholars that they do not rule
out meaningful comparison between 1951 and
subsequent Censuses provided we restrict our
analysis to the male workforce only. Because of the
varying treatment of earning dependents (to which

category a great many women workers in agriculture belong) the number of female workers tends to vary erratically. This is indeed common to a number of developing countries where reported variations in the female participation rates tend to be large. For this reason we have used figures for the male workforce only. The categories included are: (1) cultivators, (2) agricultural labourers, and (3) those working in livestock and allied activities. Adjustments have been made to make these figures comparable by following the procedures worked out by the Census of India, 1961, Paper I of 1967, Appendix II.

The Census definition of workers is, therefore, a measure of labour availability and hence of supply. Labour input, which is the relevant measure for our purpose here, can however differ considerably from labour availability.

The difference between the two in different modes of employment has an important bearing on technological choice.[5] The data on labour use in agriculture in the region are sketchy, fragmented and strictly non-comparable. Researchers elsewhere have derived time series of standardised man-days or man-hours of labour.[6] Such series cannot be computed for this region. The only studies reporting labour use data for farm business as a whole are Farm Management Studies. For our region, Farm Management Studies during the 1950s were conducted in Ferozepur and Amritsar Districts and again during 1968-69 to 1970-71 for Ferozepur only. These are based on a sample of farming households and report labour use of family workers and permanent farm labourers on both an annual and a monthly basis. As expected, they show significant seasonal variations in labour use. Labour use is classified into three broad categories of activity, namely crop production, tending of cattle, and other activities (which include both productive and social activities). Some comparative data on total yearly labour use are given in Table 2.5. The data, it may be noted, are for all workers (and not for males only) and activities include crop production and animal husbandry as well as other activities.

Table 2.5 shows a decline in the number of days worked by family workers and hired permanent labourers in Ferozepur District. For the region as

26

a whole, a decline in the demand for hired labour is
suggested also by a comparison of data from the
Agricultural Labour Enquiry 1964-65 and the Rural
Labour Enquiry 1974-75 (Table 2.6).

On the other hand, the number of agricultural
labourers also rose, not only because of natural
increase, but also due to in-migration from other
parts of India, which may not be reflected in the
Census count. Knowledge of trends in labour use
requires a deeper analysis of the factors affecting
the demand for labour. This was attempted by
Billings and Singh (1970), who calculated the
effects on labour requirements of continued
expansion in irrigated and total cropped area and of
various postulated levels of mechanisation. The

Table 2.5 AVERAGE NUMBER OF DAYS OF EMPLOYMENT   PER
          YEAR   OF   FAMI1Y   WORKERS   AND   PERMANENT
          LABOURERS IN PUNJAB
          (8 hours standard day)

	1956-57		1967-68 to 1969-70	
	Family Labour	Permanent Labourers	Family Labour	Permanent Labourers
Amritsar	239	282	–	–
Ferozepur	316	420	298	323
Region	275	357	–	–

Source: Computed from Farm Management Studies
         Reports

27

resulting projection for Punjab was a farm labour
demand in 1983-84 17 per cent lower than in 1968-69.

Table 2.6 WAGE PAID EMPLOYMENT IN AGRICULTURE OF
MEN, WOMEN AND CHILDREN AGRICULTURAL
LABOURERS, BELONGING TO AGRICULTURE
LABOUR HOUSEHOLDS,
1964-65 and 1974-75

	Numbers of full days in a year worked on average	
	1964-65	1974-75
Men	282	203
Women	173	131
Children	301	214

Note: For 1974-75, we derived the estimate for
Punjab/Haryana, as a weighted average of
figures given for the two states separately.

A more formal analytical model has been
developed by Rajkrishna (1975) . This is an
additive model with an interaction term and seeks to
explain direct and indirect changes in the use of
family labour due to introduction of irrigation,
high-yielding varieties, seeds, chemical
fertilisers, tubewells and tractors. He estimates
the model for Punjab, using various sources of data
including inter-industry input output tables, and
finds that the post-Green Revolution technologies
including mechanisation in Punjab during 1967-68 to
1972-73 have resulted in a decline in labour use per
cropped acre.
Given the number of complex factors operating
in different directions, the overall effect on
labour input is, however, difficult to predict,
especially for family labour. Generally employment
of family workers over time holding participation
rates constant would be influenced by cropping
pattern, intensity of cropping, technology of
production and post harvest technologies.[7]
A possible pattern of directions of change is
indicated in Table 2.7 which is based on personal

observation as well as a number of micro studies in this region relating to labour use, tractorisation and costs of cultivation.[8]

Table 2.7 POSSIBLE EFFECTS ON EMPLOYMENT OF FAMILY WORKERS

Source/Period	1950s to 1960s	1960s to 1970s
Cropping pattern	Small increase	Large increase
Intensity of cropping	Medium increase	Medium increase
Technology of production (mechanisation)	Small decrease	Large decrease
Post-harvest technology	Small increase	Small decrease
Overall effect on employment	Small increase	Small increase

For the region as a whole, we expect that family labour use over the last 30 years should have increased marginally, unless the leisure preferences of farming families have changed substantially over this period, for which there is little concrete evidence.

As regards hired labour, there is evidence of continuing in-migration of labour into the region from other parts of India, which suggests that the demand for such labour in Punjab continues to be strong.

4 Capital Stock and its Composition

Capital stock in agriculture is represented in this study by (i) livestock, (ii) machinery and implements, and (iii) non-residential farm buildings. The way in which these were evaluated is described in Appendix C.

The major items omitted are land improvements such as drainage, bunding and irrigation channels

29

and inventories, for example crops, seeds, fodder
and pesticides. The former had to be omitted
because of the lack of comparable data, although
land improvements that occurred through irrigation
other than canals are partly covered by 'machinery
and implements', which includes Persian wheels,
electric pump sets and oil engines with pump sets
used for irrigation. The omission of drainage,
bunding and such other items, which were relatively
more important in the 1950s, could have the effect
of exaggerating the increase of capital in the
post-Green Revolution period. Inventories were
omitted not just because of the inadequacy of data
but more fundamentally because including them in an
index of durable assets involves certain conceptual
difficulties which we wished to avoid.[9]

The growth of capital stock in Punjab
agriculture so defined is brought out by Table 2.8,
which shows that at the end of the period 1951-72 it
was higher by over three-quarters than at the
beginning. It increased throughout the period but
at a distinctly higher rate after the mid-1960s, the
percentage increase during 1965-66 to 1971-72 being
higher than that in the preceding fifteen years.

Table 2.8. GROWTH OF AGRICULTURAL CAPITAL
            STOCK IN PUNJAB (constant 1970-71
            prices)
            (Index numbers, 1951 = 100)

1951	100.00
1956	112.09
1961	122.53
1966	129.09
1972	177.73

Sources: described in Appendix C.

30

Table 2.9 COMPOSITION OF AGRICULTURAL CAPITAL
(per cent of value at 1970-71 prices)

	1951	1956	1961	1966	1972
Livestock	49.38	48.92	46.06	42.94	36.06
Agricultural implements and machinery	10.48	11.70	13.32	18.29	35.56
Farm buildings	40.14	39.38	40.62	38.77	28.38
Total	100.00	100.00	100.00	100.00	100.00

Source: See Appendix C.

Effective capacity may have grown by even more than the figures of Table 2.8 suggest, since the utilisation of capacity, especially that of livestock, improved. The small size of most operational holdings (relative to that which could be efficiently worked by a pair of bullocks) and the scarcity of custom-hiring services (due largely to the uncertainty of a timely availability) were responsible for considerable excess capacity in working cattle at the beginning of the period.[10] The rise in cropping intensity made possible by irrigation and the profitability of high-yielding varieties have encouraged small farmers to use their draught animals more intensively.[11] For tractors, too, the utilisation of capacity, as measured by hours of use per year, increased over time. This was achieved, however, mainly by extending the range of agricultural operations for which tractors were used, a trend which is reflected to some extent by the increase in various tractor-operated implements since 1966 (included in the category 'Other modern equipment').

The effective capacity of capital stock can also increase through improvements in the quality of equipment. The typical iron plough now used in Punjab is much superior to that used in the early 1950s, and the average horse power of tractors is distinctly higher. Constant price calculations, however, fail to reflect this.

It is not just the increase in aggregate capital that is of interest. The break-up of this aggregate has also changed substantially over time.

The break-up of aggregate capital as between its three broad components, viz. farm buildings, livestock, and implements and machinery, is shown in Table 2.9. During the period 1951-72, while the share of farm buildings fell from 40 per cent to 28 per cent and that of livestock from 49 per cent to 36 per cent, the share of implements and machinery rose from 10 per cent to 35 per cent. If we look at the composition of increments in capital, implements and machinery account for a little less than half of the total increase during the period, farm buildings for about a third and livestock for the remainder. The importance of investment in machinery and implements comes out clearly. However, the types of implements and machinery in use have also changed.

32

Growth of Agricultural Output and Inputs

The growth of individual items of implements and machinery and associated changes in the composition of the implements and machinery as a whole are shown in tables 2.10 and 2.11. We shall comment on some items below.
(i) Ploughs: The wooden plough, the basic implement of traditional agriculture in India,

Table 2.10 CHANGES IN COMPOSITION OF AGRICULTURAL IMPLEMENTS AND MACHINERY OF PUNJAB, 1951 to 1972

Item	1951 (%)	1956 (%)	1961 (%)	1966 (%)	1972 (%)
1. Ploughs:					
Wooden	3.73	3.73	2.85	1.75	0.67
Iron	0.66	0.90	1.26	1.46	0.71
Total	4.39	4.63	4.11	3.21	1.37
2. Carts	49.03	47.18	44.39	35.83	15.14
3. Cane-crushers:					
(i) Worked by bullock	5.60	5.25	5.02	3.94	1.39
(ii) Worked by power	0.12	0.11	0.30	0.42	0.47
Total	5.72	5.36	5.36	4.36	1.86
4. For Irrigation					
(i) Persian wheels	34.92	27.97	22.90	18.35	2.82
(ii) Oil engines with pump	0.59	1.46	1.92	4.88	9.30
(iii) Electric pump	0.12	1.96	2.05	5.38	12.78
Total	35.63	31.39	26.87	28.61	24.94
(Sub-total (ii) + (iii)	0.71	3.42	3.97	10.26	22.18)
5. Tractors	5.22	11.44	19.30	26.91	51.14
6. Other modern equipment	-	-	-	1.06	5.58
7. Other	4.16	-	-	-	-
Total	100.00	100.00	100.00	100.00	100.00

Source: See Appendix C.

33

Growth of Agricultural Output and Inputs

Table 2.11 GROWTH OF AGRICULTURAL IMPLEMENTS AND MACHINERY,
          PUNJAB, 1951 to 1972
          (Index Numbers: 1966 = 100)

Item	1951	1956	1961	1966	1972
1. Ploughs:					
Wooden	98.24	122.89	144.89	100.00	97.25
Iron	20.83	35.79	49.06	100.00	124.46
Total	63.08	76.02	82.51	100.00	109.61
2. Carts	63.22	76.15	87.72	100.00	108.49
3. Cane-crushers:					
(i) Worked by bullock	65.76	77.10	90.32	100.00	90.67
(ii) Worked by power	12.80	14.73	50.97	100.00	292.03
4. For Irrigation					
(i) Persian wheels	87.93	88.14	88.36	100.00	90.67
(ii) Oil engines with pump	5.63	17.25	27.82	100.00	170.24
(iii) Electric pump	1.02	21.07	27.07	100.00	443.44
Total	57.54	63.43	66.49	100.00	137.72
Sub-total (i) and (ii)	3.20	19.26	27.40	100.00	313.51
5. Tractors	8.97	24.58	50.77	100.00	387.10
6. Other modern equipment	-	-	-	100.00	2146.92
7. Grand total	46.21	57.83	70.80	100.00	213.68

Note: Indices of values in constant (1970-71) prices.
Source: See Appendix C.

increased substantially in number in the 1950s, declined sharply during 1961-66 and was relatively stable thereafter. The number of iron ploughs, on the other hand, increased five-fold during 1951-66 and by another 25 per cent during 1966-72. It is clear that wooden ploughs have been replaced by iron ploughs during this period, but the rate of substitution has slowed down since the Green Revolution, probably because of the advent of tractors on a large scale. This is shown, for example, by the decline in the value of all ploughs (wooden and iron) as a proportion of the total value of implements and machinery from 1956 onwards and especially since 1966.

(ii) Irrigation Equipment: While, by all accounts, increase in ground water irrigation has played a vital role in the progress of Punjab agriculture since independence, the composition of the equipment concerned has changed dramatically. Persian wheels, for example, which constituted one-third of the total value of implements and machinery at the beginning of the period, declined to just over 3 per cent of the total in 1972. By contrast, there was a very considerable increase in the use of electric pumps from 1961 onwards (the increase during 1951-56 being more a reflection of an absolutely low initial level) and particularly since the mid-1960s. Moreover, both oil engines and electric pumps increased at roughly the same rate during 1956-66. In the following period, the increase in electric pumps overtook the use of oil engines, a trend which has been maintained in the 1970s.

(iii) Tractors: Again, the sharp initial rise is perhaps of relatively little significance because of the absolutely low numbers involved. The picture begins to change in the 1960s and there is greatly accelerated use of tractors for a variety of purposes from the middle of the decade. This is also reflected in the tremendous increase in 'other modern equipment' since 1966, which is almost entirely on account of power-driven and tractor-driven equipment.

The overall picture is clearly one of increasing modernisation of implements and machinery. If 'traditional' capital is defined as consisting of wooden ploughs, carts, cane-crushers

Table 2.12   CHANGES IN COMPOSITION OF LIVESTOCK CAPITAL
IN PUNJAB, 1951 to 1972

	1951 (%)	1956 (%)	1961 (%)	1966 (%)	1972 (%)
Working animals	61.68	60.61	60.81	58.06	51.02
Milch animals	38.32	39.39	39.19	41.94	48.98
Total	100.00	100.00	100.00	100.00	100.00

Source: See Appendix C.

36

worked by bullocks and Persian wheels, and modern capital of the other items specified in Tables 2.10 and 2.11, we find that traditional capital formed about 94 per cent of the total in 1951, 61 per cent in 1966 and only about 25 per cent in 1972.

The accelerated pace of modernisation after the mid-1960s comes out even more sharply if we compare the item-by-item distribution of the increase in the value of implements and machinery during 1951-66 with the corresponding distribution in 1966-72. Of the total increase in the former period, carts accounted for about one-quarter; for 1961-66, their share of the increase was less than 3 per cent. Tractors, on the other hand, accounted for about 46 per cent of the increase in the earlier period and 68 per cent in the latter; while for electric pumps the figures are 10 per cent and 16 per cent respectively.

As regards livestock, the changes in its composition over time are shown in Table 2.12. Here the picture is fairly simple. There was some growth in the absolute level of draught animal power during the 1950s, an actual decline during 1961-66 and a slight absolute increase afterwards. Over the period as a whole, it increased marginally in magnitude. Milch animals, on the other hand, which increased at about the same rate as working animals up to 1961, showed a faster rate of growth than working animals during 1961-1966 and a much faster rate from that time. In consequence, the composition of livestock changed in favour of milch animals, a trend which has continued in the 1970s.

5 Factor Proportions, Factor Productivity and Technical Change

The analysis of growth is concerned essentially with the question: to what extent have different production factors contributed to final output? The analysis has often been done in terms of an aggregative production function, usually in its Cobb Douglas form, assuming disembodied technical change and perfectly competitive markets.

Contributions made to the observed rate of growth in real output by increases in labour and in

Table 2.13 GROWTH OF OUTPUT AND INPUTS IN PUNJAB AGRICULTURE, 1951 to 1972

	1951*	1956	1961	1966	1972
Y	100.00	125.18	173.65	184.33	278.05
Ln	100.00	105.83	109.10	105.05	112.33
La	100.00	106.05	112.96	122.31	134.80
K	100.00	112.09	122.53	129.09	177.80
Y/Ln	100.00	118.28	159.16	175.46	247.53
Y/La	100.00	118.04	153.73	150.72	206.27
Y/K	100.00	111.67	141.72	142.79	156.38
K/La	100.00	105.70	108.47	105.55	131.90
K/Ln	100.00	105.92	112.31	122.88	159.28
Ln/La	100.00	99.79	96.58	85.89	83.33

Y = Output net of material cost.
Ln = Net sown area
La = Male workforce in agriculture
K = Capital stock in agriculture =
     Agricultural implements and machinery + livestock +
     farm buildings.

* Note: For details of 1951 figures see Appendix B

Source: Computed from output and input series described in Tables 2.1
        to 2.12 above.

capital are measured by multiplying the growth rates of the input concerned by its cost share. The excess, if any, of the observed rate of growth of output over the sum of contributions from labour and capital is then regarded as a measure of technical progress. On the basis of such studies it was found that the residual technical progress accounted for most of the overall growth of output in the USA and elsewhere.[12] A number of empirical studies have applied this method in analysing the growth of agricultural output, the only change being the inclusion of land as an additional factor of production.[13]

Such an analysis of growth involves a number of serious difficulties. The existence of an aggregate production function requires not only that each productive factor can be represented by a single number, but also that the aggregation involved should be meaningful in terms of the analytical framework itself. The conditions required for such meaningful aggregation to be possible, especially in the case of capital, are extremely stringent.

Estimates based on such an exercise may therefore be conceptually invalid except in special circumstances. Moreover, the value measure of capital used in aggregate production function analysis is not in general independent of prices and distribution. Hence, using the factor shares to measure the contribution of capital involves circular reasoning.[14]

As far as empirical analysis is concerned, the basic difficulty lies in interpreting the residual as a measure of technical progress. The measured residual in a statistical analysis will include much else, viz. omitted variables as well as errors of measurement. Such errors may simply be errors of observation. They may also be specification errors, arising from the fact that the data used may not correspond to concepts implied by the theory. In the present context, the differences in the intensity of labour used, which we have discussed above, are an important example. Specification errors may also arise from complementarity of capital and other productive factors. It is often through new additions to capital stock that technical change is introduced into the economic system, especially if the system concerned is one of

Growth of Agricultural Output and Inputs

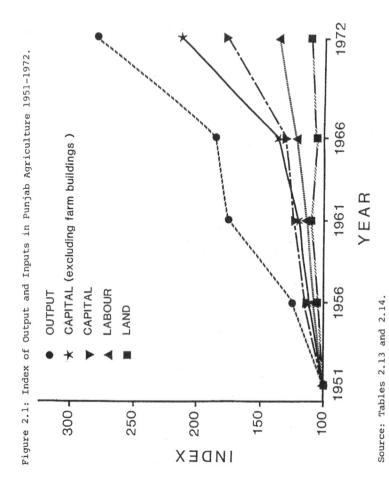

Figure 2.1: Index of Output and Inputs in Punjab Agriculture 1951-1972.

Source: Tables 2.13 and 2.14.

traditional agriculture. Technical change is then said to be 'embodied'. A rise in the rate of investment, a decline in the average age of equipment and a faster rate of technical progress thus tend to be interrelated. The importance of embodied technical change in the Punjab context has already been brought out by our discussion.

For these reasons we have not followed this approach. Instead we confine ourselves to presenting broad trends in factor productivity and factor intensity during the period under review and draw some overall conclusions.

Table 2.13 presents the indices of value added in agriculture (Y), labour use (La), net sown area (Ln) and capital stock (K) at five-yearly intervals since 1950-51 and basic ratios representing output per hectare (Y/Ln), output per worker (Y/La), capital per worker (K/La) and capital per hectare (K/Ln). Trends of these indices and ratios are shown in figure 2.1.

The table shows a substantial rise in output over the period. Inputs, namely net sown area, male workforce in agriculture and fixed capital, also increased, but by much less than output. There has been a spectacular rise in output per hectare, a substantial rise in output per worker, and a significant rise in output per unit of fixed capital. The rise in all three ratios unmistakably represents rising efficiency of production, i.e. an increase in 'total factor productivity'.

Factor intensities for the period as a whole have changed. As can be expected in a land-scarce situation, with the labour force in agriculture still increasing, output per net sown hectare rose faster than output per worker. A rising intensity of land use partly reflects this.[15]

In our analysis of growth rates at the start of this chapter, we divided the period 1951-72 into a number of sub-periods partly because the agricultural policy had been changing over time and partly because measured growth rates in different sub-periods do in fact turn out to be significantly different. That changes in the levels of various inputs and in their productivity followed a temporal pattern roughly similar to that observed in the growth rates of agricultural output is indicated by Table 2.13.

The decade of the 1950s saw a major extension of irrigation, the consolidation of land holdings and improvements in institutions serving agriculture. Value added (Y), net sown area (Ln), workforce in agriculture (La) and capital (K) increased, though by different magnitudes. The average productivity of land (Y/Ln) and that of labour (Y/La) increased at roughly the same rate while capital productivity (Y/K) rose less. The land-labour ratio declined slightly while capital per worker (K/La) and capital per hectare (K/Ln) both increased. The overall picture is one of considerable increase in the efficiency of production and a slight shift in factor proportions.

A comparison of 1966 with 1961 could be biased by the abnormal weather of 1965-66. This should have affected land area (Ln) and output (Y) in 1966. The impact on capital (K) is likely to be marginal and labour (La), given our measure of labour, should not be affected at all. Increases in output (Y) and capital (K) during 1961-66 were small; net sown area (Ln) actually declined. Output per hectare (Y/Ln) continued to increase (perhaps because of the lagged effects of public irrigation during the previous decade). Output per worker (Y/La) declined and output per unit of capital (Y/K) hardly changed. Capital per worker (K/La) and land per worker (Ln/La) declined. The overall picture is indicative of the setting in of diminishing returns in the agricultural sector.

During 1966-72, there is a sharp acceleration in the growth both of net value added (Y) and of capital (K). The change in labour (La) was similar to that in earlier periods while that in net sown area (Ln) registered a slight increase. The slow increase in area as compared to the 1950s seems to be the result of possibilities of extending the net sown area being nearly exhausted.[16] Average productivities of land, labour and capital taken individually increased during this period though at different rates, that of land rising the most. Again, an increase in the efficiency of production must have taken place.

During this period factor proportions also changed significantly. Land per worker declined further while capital (both per worker and per hectare) increased. Despite claims about the biological and land-saving nature of the Green Revolution technologies, capital thus seems to be replacing labour and land at the margin. In this context we would like to remind the reader that our capital series represent durable items, viz. farm machinery and implements, farm buildings and livestock: they exclude single-use items such as chemical fertilisers, fuel and other materials, and output is also measured net of these.

So far, we have been discussing the results of Table 2.13 in which capital stock includes (i) livestock, (ii) implements and machinery and (iii) farm buildings. It may also be useful to use a concept of capital consisting of directly productive items only, i.e. exclusive of farm buildings. Let us denote this as K'. The corresponding indices, i.e. K', Y/K', K'/La and K'/Ln, are given in Table 2.14. The other indices are not affected and remain the same as in the previous table.

A comparison of Tables 2.13 and 2.14 shows that up to 1961 the rate of growth of capital is hardly affected by whether farm buildings are included or not. Between 1961 and 1966, K' increases somewhat more than K, and between 1966 and 1972 much more. The result is that, during 1961-66, Y/K' shows a small decline rather than the small rise recorded by Y/K while between 1966 and 1972 Y/K' shows a clear decline as against an increase in Y/K. Similarly, for the period following the Green Revolution, i.e. between 1966 and 1972, the rates of increase in K'/Ln and K'/La are much higher than in K/Ln and K/La respectively. Between 1961 and 1966 they are slightly higher, and for the decade of the 1950s there is hardly any difference. The point is not simply that different measures of capital led to different results, although this is trivially true. It is that after the mid-1960s, the pace of growth of capital stock was set essentially by investment in directly productive capital, especially that in implements and machinery: hence a measure of capital - such as K' - which focuses on this aspect would show a greater rate of rise in capital, both absolutely and relatively to output or to other inputs. Changes in factor intensities reflect changes in 'techniques' or in 'technology'. Ishikawa (1981), in distinguishing 'technique' and 'technology', suggests that four interconnected

43

Table 2.14   GROWTH OF OUTPUT AND INPUTS IN PUNJAB
1951 - 1972

	1951	1956	1961	1966	1972
Y	100.00	125.18	173.65	184.33	278.05
K	100.00	112.09	122.53	129.09	177.80
Y/K	100.00	111.27	141.72	142.79	156.38
K/La	100.00	105.70	108.47	105.55	131.90
K'	100.00	113.52	121.55	132.04	212.73
Y/K'	100.00	110.17	142.86	139.60	130.71
K'/La	100.00	107.05	107.60	107.96	157.81

Notes: Y   = Net Value Added
         Ln = Net Sown Area
         La = Male Workforce in Agriculture
         K   = Agricultural implements, farm machinery, milk and
               draught animals and farm buildings.

         K' = K exluding Farm Buildings.

44

components of a technology are significant in making it unique. These are:
(i) labour and other current inputs,
(ii) land and its structure,
(iii) other capital assets,
(iv) farming methods.
Using his classification we may expect to find a number of technologies coexisting in our region. The critical element which distinguishes these technologies from one another is (iii), namely capital assets. Introduction of a new capital asset, e.g. a tractor, would not only affect land preparation but a whole range of farm practices and activities, thus affecting each of the other three elements (i), (ii), and (iv). Similarly, the introduction of a tubewell would affect the use of labour and other current inputs even when the crop mix and the seed variety used remain unchanged. But we do know that these also changed.

It is well known that farm practices evolve around a crop mix, crop rotation and crop calendar.[17] Historical experience of Europe, USA and Japan regarding the introduction of farm machinery clearly demonstrates that single operation farm machines rarely succeed and introduction of any major agricultural machinery necessitates readjustment in crop mix, farm practices and total farm business operations.[18] Thus farm technology is an integrated whole and a change in some parts will necessitate a readjustment of the total farm business operation.

The commonly held belief that the Green Revolution technologies are scale neutral and that by deliberate policy their adoption can be made mainly land saving, leaving the overall technological mix unaffected, ignores the complexities of the interconnectedness of these technologies. In a market-based agriculture, economic policy cannot, without drastic interventions, arrest or reverse the emerging trend of capital deepening. In fact this trend opens up possibilities of major structural change which seem to be appearing already. Policy, however, can influence research and development, working of the market, and income distribution, which have important implications for further growth and for the nature of development in this region.

NOTES

1. 'Precise uses of "growth rates" are
entirely inadmissable whether for comparing
different countries or short periods of the same
country. Their computation is largely arbitrary'
(Oskar Morgenstern, 1963, p.300).
2. The extent of variation in rainfall can be
seen in Table 3.1.
3. See Appendix C tables for input data
series.
4. A good discussion is in Giri (1962).
Trends in land use at the all-India level up to
1963-64 are discussed in Giri (1966).
5. See, for example, Sen (1975), Ishikawa
(1981) and Johnston (1966).
6. Denison (1962), Ishikawa (1981), Sawada
(1979), and Hayami and Ruttan (1971).
7. Ishikawa (1981) explores the implications
of changes in technology and crop intensity in the
context of Japanese and Chinese historical
experience which is mainly concerned with rice
production.
8. See, for example, Kahlon and Singh (1973),
Hanumantha Rao (1979), Agarwal (1980).
9. See Sen (1964).
10. See Sharma (1982).
11. Sharma's(1982) study of Haryana finds that
the large farmers during 1961-72 acquired tractors
to meet their increased demand for draught power;
medium farmers added draught animals or, sometimes,
hired tractor services; small farmers managed to
meet their additional requirements by utilising the
idle capacity of draught animals they already owned.
12. See, for example, Solow, R.M., 'Technical
Change and the Aggregate Production Function', in
Review of Economics and Statistics, Vol. 39, 1957.
13. See, for example, Hayami, Yujio,
A Century of Agricultural Growth in Japan, Its
Relevance to Asian Development, University of
Tokyo Press, 1980. The conclusions reached have
been broadly similar to the earlier studies.
14. Surveys of the literature are given in G.C.
Harcourt, 'Some Cambridge Controversies in the
Theory of Capital', Journal of Economic Literature,
Vol. 7, 1969; and in M.I. Nadiri, 'Some
Approaches to the Theory and Measurement of Total

Factor Productivity: A Survey',
Journal of Economic Literature, December 1970. A
lucid presentation of the outcome of the debate is
given by Hicks in his Nobel lecture, 'The Mainspring
of Economic Growth', published in Swedish Journal of
Economics, December 1981. Alternative econometric
approaches are discussed in K.D. Patterson and
Kerry Schott, The Measurement of Capital,
Theory and Practice, London, MacMillan, 1979.
        15. See Table A.2.
        16. See our discussion of land use above.
        17. See, for example, ICAR (1961) and Singh,
Johl, et al. (1966) for farm practices, farm
technology and crop calendars in this region and for
farming systems around the world.
        18. See Heady (1951) and Jones (1954) for
implications of mechanisation in USA and UK
respectively.

Chapter III

INFRASTRUCTURE AND THE ROLE OF THE STATE

*Thus, the productive forces are the result of man's practical
energy, but that energy is in turn, circumscribed by the
conditions in which man is placed by the productive forces
already acquired, by the pre-existing form of society, which
he does not create, which is the product of the preceding
generation.*

Karl Marx

The infrastructure component of public investment
has been recognised as an important element in the
process of economic development at least since Adam
Smith, who described the provision of infrastructure
as one of the three important duties of the
sovereign. Erecting and maintaining public
institutions and public works was, he wrote, the
'third and last duty of the sovereign', the first
two being defence and justice. The rationale of the
sovereign doing so, according to Smith, was that
such works and institutions, though they might be
'in the highest degree advantageous to a great
society', were of such a nature that the profit could
never repay the expense to any individual or small
number of individuals', and it, therefore, cannot be
expected that any individual or small number of
individuals should erect or maintain them (Smith,
1776, p.214).

The role that such infrastructure has played in
the recent economic development of Punjab is briefly
described in the present chapter. The chapter
consists of nine sections. The first distinguishes
between three broad patterns that investment in
infrastructure could follow. The next two deal with
investment in irrigation and power generation
respectively. The fourth describes the institutions
for rural credit and marketing facilities and the
fifth those concerned with Agricultural Research and
Development as well as Agricultural Extension
Services. Improvements to the road transport
system, health and education, and urban services are
described in the next three sections. The last

48

describes how investment in infrastructure was financed.

1 The Development of Infrastructure

Modern development economists, while recognising the importance of infrastructure, have been rather ambiguous about its content, Youngson's (1967) being one of the few illuminating contributions on the subject. Infrastructure, by common consent, includes all things provided by the state which promote directly productive activities. Transport, irrigation, power, water supply, health, education and urban services are usually considered as important elements of infrastructure. The aspects of investment in infrastructure that have been emphasised in the recent literature are the externalities that it involves, its high initial cost and long period of gestation and, in a great many cases, the fact that marginal cost tends to fall as output increases. These create special problems in equating demand and supply of infrastructure if its provision is left to the market, even in a market economy. In a mixed economy with a relatively important public sector, the provision of infrastructure tends to be planned and provided by the state.

What kind of strategies for providing infrastructure are followed by the state will depend upon its objectives and on prevailing circumstances. Thus a colonial government's objective in building infrastructure, as pointed out by Lewis (1978), Birnberg and Resnick (1975) and many others, is usually to create and extract surplus which invariably gives rise to a colonial pattern of development. More generally, any state dominated by a particular class or social group could be pressurised into creating infrastructure that would provide maximum economic opportunities for that class or social group.

In a planned economy based on balanced growth, the conflict of class interests can be softened to some extent in the early stages of growth by promoting the cause of the economic development of the region as a whole, which could be speeded up by building infrastructure ahead of demand.

Infrastructure and the Role of the State

Hence this was the aspect of investment in infrastructure which exponents of the 'balanced growth' approach to planned economic development, notably Nurkse (1952 and 1961), emphasised. Nurkse regarded overhead or infrastructural investment as an essential framework for miscellaneous economic activity, as a 'non-specific, initiatory pioneering type of investment' (Nurkse, 1961, p.75), which provides inducements for directly productive investments and hence tends eventually to create its own demand. On this view, the nature of investment in overhead capital is such that, unlike other kinds of investment, it is undertaken not to meet a need that already exists but rather to create one. Such a view also implies that an overhead capital structure built in an underdeveloped country will not initially have enough demand to ensure that its capacity is fully utilised. Its justification is of a long-term nature.

As against the view that infrastructure should be built ahead of demand, exponents of 'unbalanced growth' such as Hirschman (1958) favour a strategy of creating infrastructure only in response to demand for it. Hirschman reasons that building infrastructure ahead of demand will lead to wasteful uses of capital. On the other hand, if the demand for infrastructure is allowed to develop first, planning problems will be simplified because pressures felt by public authorities 'to do something' will act as a powerful incentive. Since, argues Hirschman, the desire for political survival is at least as strong a motive as the desire to make a profit, we may in such circumstances expect some effective action to be taken.

In the process of economic development of Punjab the investment in infrastructure by the state may be said to have followed all three patterns: viz. colonial, ahead of demand, and in response to it, patterns being different in different periods in its history and for different components of infrastructure. This had rather interesting influences on private investment activity and the pattern of economic development that followed. We give below a telescopic view of the evolution of irrigation, transport, power, health and education sub-systems including the interaction between public and private investment.

50

2 Public Investment in Irrigation

A crucial condition for settled agriculture is the availability of sufficient moisture in the soil to make plant growth possible. As was noted in Chapter I, in both Indian and Pakistani Punjab not only is rainfall low but it also has a large regional and temporal variation, as is clear from Table 3.1.

Table 3.1 RAINFALL IN TWO PUNJABS (inches)

Rain Belt	District	1952-56	1957-60	1961-64
High belt	Pindi	37.4	48.7	n.a
	Gurdaspur	41.2	54.8	46.3
Medium belt	Lahore	18.9	22.3	23.4
	Amritsar	21.3	26.2	24.9
Low belt	Multan	6.5	6.6	7.4
	Ferozpur	14.1	16.1	15.6

Note: The districts are listed in pairs, with the first district in each pair belonging to Pakistan Punjab.
Source: For Indian Punjab see Brown, 1967: p.2; for Pakistan Punjab, GOP, District Census Reports, 1961,Census; GOP, Year Book of Agriculture Statistics, 1971-72.

Such a pattern of rainfall which, moreover, is concentrated in only three months of the year, severely restricted the crop calendar. Intensive land use over the whole year would have been impossible without provision of artificial irrigation. The region had known well (lift) irrigation from times immemorial in regions where the water table was high enough to allow for lift irrigation with the use of Persian wheels drawn by a pair of bullocks. In the pre-British period agricultural development in Punjab was concentrated in these regions only.
Following the uprising of 1857, the British colonial administrators saw the special potential of Punjab either as a source of surplus and support or

as a source of serious trouble. The newly developed
European technology of flow irrigation with the use
of gravity dams was ideally suited for the
development of sparsely populated regions which had
a precarious food-population balance because of a
short and unstable crop calendar. The regions in
Punjab which were developed with the help of this
technology are now called Canal Colonies and are
part of Pakistan. For these considerations Punjab
obtained a large chunk of public investment in
irrigation in British India during 1860-1947 as is
clear from Table 3.2.

Table 3.2 PUBLIC INVESTMENT IN IRRIGATION:
1860-1947
(Rs. Million)

Period	Punjab	India
1860/61 to 1897/98	135 (24.2)	558
1898/99 to 1918/19	315 (47.0)	669
1919/20 to 1946/47	590 (30.0)	1968

Note: Figures in parenthesis are percentages of
total all India figures.
Source: Thavraj (1963).

The consequences of this regional pattern of
public investment in irrigation were in accordance
with the colonial administration's expectations.
The desired surplus for export was obtained from a
few districts only and these districts also became a
very dependable source of recruitment for British
India's armed forces. Despite such pockets of
prosperity the rest of Punjab was unaffected by this
development as indeed is typical of a colonial
pattern of development (cf. the Royal Commission on
Agriculture, 1928; Darling, 1925; and Mushraff,
1980). On the eve of the partition of India in
1947, 80 per cent of canal-irrigated areas and
surplus-food producing regions went to Pakistan as
is clear from Table 3.3.
The partition of India and its aftermath,

Infrastructure and the Role of the State

Table 3.3 IRRIGATED AREA, TWO PUNJABS, 1944-45
(Million Acres)

	Canal Irrigated Area	Wells and Tubewells* Irrigated Area	Total** Irrigated Area
Pakistan Punjab	10.15	2.08	12.51
Indian Punjab	2.93	2.02	5.04
Total Undivided Punjab	13.08	4.10	17.55

Notes: * This is almost wholly wells (as tubewells
were extremely few at the time).
** Includes, apart from canal and wells, also
tank and other sources.
Source: Musharaff(1980)

Table 3.4 IRRIGATION IN PUNJAB ACCORDING TO SOURCE
(% of Net Area Irrigated)

Source	1950-51	1960-61	1965-66	1970-71	1975-76
Canals (govt.)	9.10	58.07	56.92	44.93	43.80
Canals (private)	n.a.	0.35	0.27	0.21	0.13
Wells	84.29	36.68	31.98	15.30	
Tubewells	n.a.	4.36	8.71	39.79	55.85
Others	6.61	0.54	2.12	0.17	0.22
	100.00	100.00	100.00	100.00	100.00
Net area irrigated as % of net area sown	20.5	54	59	71	75
Gross area irrigated as % of GCA	22.0	56	64	75	79
All India (% of Net Area Irrigated)	17.1	18.5	19.9	23.3	29.9

Sources: Statistical Abstracts of Punjab and Haryana
(various issues).

described in Chapter I, forced the national policy makers and the Punjab government to accord the highest priority to the Bhakra Nangal project, which had been in the planning stage since 1946. This project was expected to provide canal irrigation, electric power and flood control.

The results turned out to be even better than the policy makers themselves had expected. Because of refugee resettlement, the consolidation of land holdings and the construction of irrigation channels and of feeder canals were facilitated. Since a large proportion of the refugees had previously been residents of the Canal Colonies of West Punjab, they were used to wet irrigation through gravity canals. The proportion of area irrigated in Punjab and Haryana increased substantially and the rate of utilisation of irrigation facilities was also high. The break-up of irrigation according to source is reported in Table 3.4.

Clearly, canal irrigation provided by the state as part of the infrastructure continued to be a dominant source of irrigation till 1965-66 when private wells and tubewells became relatively more important.

This shift caused by exigencies of the Green Revolution, timely and controlled irrigation being required for the success of high-yielding varieties of seed, has postponed the problem of drainage and water logging currently considered to be serious in Pakistani Punjab. The profitability of private investment in tubewells was due to the successful adoption of HYV seed but the state facilitated such investment by making more rural credit available and by its own investment in rural electrification programmes.

3 Power Generation and the Agricultural Sector

The Bhakra Nangal project was a multipurpose project. In addition to irrigation and flood control, it was expected to provide hydroelectric power. Electricity power generation in Punjab increased from 170 million kilowatt hours in 1947 to 1819 million kilowatt hours in 1965-66, a more than 10-fold increase. During the decade 1965-66 to 1975-76 power generated doubled, as is clear from

Table 3.5.

Some 43 per cent of power generated in 1975-76 was used for tubewell irrigation. This lends credence to our statement above (see Table 3.4) that the expansion of irrigation in Punjab after 1965-66 was largely due to private investment in electric tubewells made possible by the availability of power.[1] By 1975-76, 100 per cent of the villages in Punjab had been electrified and per capita power consumption in the region was more than double the all-India average of 100 kW.h. Punjab does not have any major large-scale industry nor does it have a big city like Bombay or Delhi. A substantial part

Table 3.5 DEVELOPMENT OF INFRASTRUCTURE AND POWER GENERATION IN PUNJAB

	Punjab			India
	1947	1965-66	1975-76	1975-76
Capacity (MW)	48	432	1009	20,127
Power Generated (m. kW.h)	170	1819	3841	79,230
Power Availability in 10 kW.h per 10 of population	–	149	262	131
Power Consumption per capita	–	N.A.	231	110
Per Capita Power Consumption for Irrigation (kW.h)	–	–	61	15
% of Villages Electrified	–	–	100	–

Note: Irrigation accounts for 43% of power consumed in the state.
Source: G.S. Bhalla, The Green Revolution in the Punjab (India), mimeo, JNU, 1982.

of the electricity consumed is in the agricultural
sector. Per capita power consumption for irrigation
in Punjab is four times the all-India average of 15
kW.h.. Still, electrification of rural Punjab in
itself might not have led to private investment in
tubewells on the scale that occurred, if the state
had not taken the complementary step of expanding
the availability of rural credit through cooperative
credit arrangements.

4 Rural Credit and Marketing Institutions

Cooperative rural credit in Punjab has a long
history. The Royal Commission on Agriculture (1928)
had recommended the establishment of primary
agricultural cooperative societies (PACS) in the
state and the British administration took some steps
in this direction during the 1930s. A major boost
to the movement was given by the government of
Punjab after 1947. By 1971 all villages were
covered by PACS and membership had risen to 1.5
million farmers which accounts for 98 per cent of
the farming households in the region. Almost all
societies were active and 75 per cent of the members
borrowed from the societies.[2] The supply of rural
credit was augmented through Land Development Banks
(LDB) and Taccavi loans as well. Details of rural
credit availability during 1961-74 are given in
Table 3.6.
     Taccavi loans are emergency loans for
short-term purposes and go up in periods of drought
and go down to negligible levels in periods of
bumper harvest. Therefore for purposes of private
investment in agriculture loans from LDBs and PACS
are more relevant. In 1970-71 51 per cent of the
long term loans were for installation of
tubewells.[3]
     Although the overall record of expansion in
rural credit is impressive, it conceals large
variations among different farm size groups. As
could be expected, large farmers benefited from
these institutional sources of cheap credit with
average nominal rates of interest of about 9-10 per
cent in a situation where the inflation rate was
about 10 per cent, much more than the small farmers.
These details are discussed in Chapter V.

Infrastructure and the Role of the State

The state also attempted to make the regulation of agricultural marketing more effective by revising the Market Committee Act of 1938 and through the introduction of a Market Intelligence Service in 1957. Through the Warehousing Corporation, storage facilities were substantially expanded in the decade of the 1960s which made the handling of HYV wheat production feasible. Cooperative marketing societies and state trading in food grains provided the necessary competition as well as diversity for

Table 3.6 INSTITUTIONAL AGRICULTURAL CREDIT SUPPLY, BY SOURCE IN PUNJAB

	LDB *	PACS **	Taccavi ***	Total
		(in million rupees)		
1960-61		118.0		
1965-66	17.5	275.5	107.3	400.3
67	16.5	248.9	47.8	313.2
68	50.5	329.5	24.9	404.9
69	153.0	578.1	5.1	716.2
70	178.9	528.1	7.7	714.7
71	195.7	572.7	43.5	811.9
72				
73		620.0		
74		590.0		

Notes: * Land Development Bank
      ** Primary Agricultural Credit Societies
      *** Taccavi loans are state loans disbursed through the Revenue Department in Punjab.

      Ordinary commercial banks also advance agricultural credit in Punjab but they are not dominant sources and data for them are not available.
      PACS loans in Punjab do not include the long-term loans advanced by these cooperative societies.

Sources: Sen and Amjad (1977, p.41); Etienne (1976, p.19).

marketing and storage arrangements in the region.
Farm Management Studies of 1968-71 for this region
suggest that all classes of farm households
participated in these facilities but large farmers,
as usual, benefited more than others from the
created infrastructure.

5 Research and Development and Agricultural Support
Institutions

In the 1950s the government of Punjab launched a
number of programmes to improve Research and
Development and agricultural extension services in
the state. Special programmes for farmers were
introduced as part of the All India Radio broadcasts
from Jullundhur and Amritsar and, as a matter of
routine, they reported farm prices in all the major
markets of Punjab. As part of the national
Community Development Programme all village
Panchayats had acquired radio sets and 99 per cent
of these in Punjab villages were in regular use.
Details of improved farm practices and other
relevant farming information were also relayed
regularly.

In 1957 the Punjab Agricultural College was
upgraded to the status of a university and was given
responsibility for agricultural extension in nine
districts of the state. The newly created Punjab
Agricultural University promptly established a large
number of experimental stations in different parts
of the state with laboratories equipped with soil
testing facilities. The University Council was
constituted with a majority of Council seats held by
farmers, inevitably large farmers. This forged a
major link between R and D in agriculture and its
user community. Thus when high-yielding varieties
of wheat became available in the mid-1960s, the
preconditions for their quick adoption had already
been established.[4]

6 Roads and Transport Development

Road development programmes of the British colonial
administration, like its irrigation programme, were
motivated partly by a desire to extract agricultural

surplus from the Canal Colonies in West Punjab (now
Pakistan) and partly by administrative purposes.
The share of total road construction-investment in
India allocated to undivided Punjab was substantial,
as is clear from Table 3.7, but it was concentrated
in West Punjab.

A major programme for improvement of national
highways was started by the central government
through CPWD (Central Public Works Department) in
1950. As part of this programme the Grand Trunk
Road, which had been in existence in some form or
other since the thirteenth century, was widened and
upgraded. The State government's Public Works
Department (PWD) was also reorganised and expanded
in the mid-1950s and a major programme of feeder
road development was launched during the Kairon
administration of the late 1950s. As part of this
programme the slogan was to link every village to
the nearest market and town and every town with
district and state headquarters. A number of
innovative financing arrangements for road
development were also introduced. Villagers
requesting for a feeder road were expected to supply
labour free while the state PWD would provide
materials, skills and other inputs. Thus by the mid

Table 3.7 PUBLIC INVESTMENT IN ROADS: 1860-1947
(Rs in Million)

Period	Punjab	India
1860/61 to 1897/98	92 (13.3)	691
1898/99 to 1918/19	103 (11.7)	879
1919/20 to 1946/47	270 (12.1)	2237
1860 to 1947	465 (12.2)	3807

Note: Figures in parenthesis represent percentage
      of all-India total.
Source: Verma (1980, p. 17).

1970s every village in the region was connected to a
market town by an all-weather road. The spectacular
increase in road mileage that occurred can be seen
from Table 3.8. Particularly notable was the
acceleration of road development after 1970,
resulting in a doubling of metalled road mileage in
less than five years.

The expansion of road mileage after 1965-66
consisted largely of feeder roads and seems to have
been of the kind suggested by Hirschman, i.e. in
response to demand. A vastly expanded road network
and greater marketed surplus seem to have stimulated
private investment in goods transport by road as is
clear from the substantial increase in number of
registered private vehicles in Punjab during the
1960s and 1970s. Mobility increased both due to the
introduction and expansion of public bus transport
and to the use of motor cycles and tractor trollies
on private account. However, the increase in the
number of cars has been relatively small.

7 Health and Other Merit Goods

Public investment in health, education and other

Table 3.8 METALLED ROADS MILEAGE AND DENSITY IN
PUNJAB

YEAR	total length (miles)	DENSITY	
		Miles per 100 sq. miles	Per 100,000 population *
1966	3,752	19.2	27.8
1970	6,200	31.8	45.9
1974	12,181	62.5	90.2

Note: * taking 1971 population for Indian Punjab.
Sources: Sen and Amjad (1977, p.22); Etienne
(1976, p.11); Randhawa (1974, Ch.5); Day
and Singh (1977, Ch.8).

'merit' goods has been slowly increasing since 1950 but the increase seems to have accelerated since 1970. This partly reflects a rising demand for these services due to increasing prosperity, but partly also the state's desire and ability to provide such goods. By the mid-1970s each village was served by a primary health centre, a primary school and a trained midwife.

The region has the highest retention rate in primary schools and the largest proportion of primary school leavers going to secondary schools.[5] The demographic implications of this are discussed in the next chapter.

8 Expansion of Educational Facilities

The urbanisation of this region seems to have been moving somewhat more slowly than in the rest of India. Since 1951 the rate of growth of big cities in Punjab has been substantially slower than in the rest of India although medium-sized towns have grown faster than in the rest of the country.[6] This may be partly due to the immediate effects of a fast-growing agricultural sector but state policies of promoting growth centres pursued actively since the mid-1960s may have also contributed to the outcome. The provision of infrastructure, particularly electricity, roads, education and health facilities, has also succeeded in reducing rural-urban differences to some extent, contrary to what is usually asserted by exponents of an 'urban bias'.[7]

The nature of public administration in Punjab has become 'development' oriented and the crime statistics indicate a major shift in the nature of crimes committed. Major killers in Punjab after natural causes and illness are 'accidents' and major crimes are petty economic crimes.

9 Public Revenue and Expenditure

The development orientation of public administration in this region required a major tax effort on the part of the state. The success of this effort was quite remarkable especially when compared with other states of India. The expansion of economic activity itself increased tax revenue and hence public expenditure could keep pace with the needs of the region. The growth of tax collection since 1968-69 is shown below in index number form, with 1968-69 = 100.

Year	Tax Collection[8]
1968-69	100
1969-70	144
1973-74	198
1974-75	213

In the Indian federal system states do not have the authority to impose income tax. The share of central revenues going to the states is governed by constitutionally established Finance Commissions every five years. During the 1950s and 1960s Punjab, because of its special and massive needs of rehabilitation, obtained more than its normal share of central funds. That a large share of public expenditure in the region was allocated to the agricultural sector is clear from Table 3.9.

Table 3.9 PUBLIC SECTOR EXPENDITURE ON AGRICULTURE

| Year | Expenditure on Agriculture * | |
| | Haryana | Punjab |
	% of total expenditure	% of total expenditure
1967-68	77.8	75.0
1968-69	69.7	73.0
1969-70	78.3	77.5
1970-71	67.8	70.0

Note: * includes agriculture, cooperation, community development, irrigation, flood control and power.

Sources: Statistical Abstracts of Haryana and Punjab, various issues; Commerce, 1972, p.71.

In a sense, public administration in the region has been oriented towards rural and agricultural development. The needs of industrialisation have now to be foreseen in the same way as needs of agriculture were anticipated during the 1950s and 1960s. Some of the implications of this will be discussed in subsequent chapters.

NOTES

1. The number of private tubewells increased in the region from 4.3 thousand in 1956-57 to 11.8 thousand in 1965-66 and to 179.7 thousand in 1974-75. See Statistical Abstracts of Punjab and Haryana, various issues.

2. For details see Randhawa (1974, p.89), GOI (1967, p.34) and Commerce (1972, pp.29-31).

3. Part of the credit for this goes to the well-informed Chief Minister of Punjab of that period, Mr Partap Singh Kairon, who himself belonged to a prosperous farming family and had been exposed

to ideas on agricultural modernisation in Europe and USA.

4. For details see Chaudhri(1979).

5. For details see AERC(1970) and Chaudhri(1979).

6. Source: <u>Statistical Abstracts of Punjab and Haryana</u>.

7. See Lipton (1977) for a cogent view of urban bias and Mitra (1977) for reasoning as to why such a view may be misinformed.

8. Refers to Punjab state only and is reported in the <u>Statistical Abstract of Punjab</u>, 1980 issue.

Chapter IV

DEMOGRAPHIC TRANSITION IN PUNJAB

*An abstract law of population exists for plants and animals
only, and only in so far as man has not interfered with them.*

Karl Marx

In our analysis of the growth of agricultural output
and inputs in Chapter II, the agricultural labour
force was treated as exogenously given and our
measure of labour supply recorded those 'dependent'
on agriculture for their livelihood. In this
chapter we shall analyse trends of population change
and the interaction between economic growth and
demographic change. The chapter is divided into six
sections. The first deals with the relationship
between economic growth and population change. The
second reports trends of population change in
Punjab. The third section is devoted to vital
statistics and migration and the fourth to
dependency ratios and household compositions. The
fifth section deals with children's participation in
the labour force and the last examines changes in
the quality of the labour force and its educational
characteristics.

1 Economic Growth and Population Change

The interaction between population and economic
growth has been examined by scholars at least since
the time of Malthus. While some important insights
have been gained, considerable difference of opinion
on the subject still persists.[1]
    In the Malthusian approach, the growth of
population is regarded as an outcome of a race
between 'the passion between the sexes' and the
means of subsistence. An increase in real income
per head has the effect of weakening the constraint

imposed by the subsistence factor, thus encouraging higher fertility; and it also raises the level of consumption, specially of food, which improves nutrition, reduces susceptibility to pestilence and disease, and brings about a decline in mortality. An increase in population results, which in turn tends to depress the standard of living. Both population and per capita real income thus get stabilised together in a low-level equilibrium trap from which the only escape is through drastic measures of population control.

Neo-Malthusian theories have not only become part of the received doctrine of development economics, they also explain the overriding importance given to family planning by international agencies such as the World Bank, which in turn has influenced population policy in less developed countries such as India.[2]

A somewhat different perspective on the population problem of developing countries, which emphasises that the response of fertility and mortality to changes in income depends on the stage of socio-economic development, is provided by the theory of demographic transition. This attempts to generalise observed population changes in economically advanced countries by postulating the following sequence of demographic patterns:

(i) a high birth rate accompanied by a high but fluctuating death rate leading to slow-growing population, interrupted by 'bad' years when population actually declines;

(ii) a high birth rate and a low death rate, leading to population 'explosion';

(iii) very low death rates and birth rates, leading again to slow-growing population.

The first stage is regarded as characteristic of pre-industrial societies and the third of economically advanced countries in their mature stage. The imbalance between birth and death rates is essentially a characteristic of the transition phase which according to some recent writers can be shortened by accelerating economic development through policy measures.

A more recent approach to the analysis of population growth is the so-called New Household Economics, which concentrates on changes in fertility. These are seen, following the neoclassical tradition, as the outcome of rational economic decisions on the part of individual households. A cogent presentation of this approach was given by Schultz (1980) in his Nobel lecture.

Only a few writers have stressed favourable effects of population growth. Among them are Hicks (1957) who sees the Industrial Revolution itself as a 'vast secular boom brought about by the unprecedented rise in population', Simon (1980) who describes in some historical detail the scale economies associated with large populations, and Boserup (1965, 1981) who rejects the pessimism of Malthus and Ricardo about agriculture's ability to feed the growing population due to the law of diminishing returns, asserting on the contrary that the pressure of population on the land has served as the principal inducement to technological innovation.[3]

Though none of the theories mentioned can provide an adequate explanation of population behaviour, they do suggest some relevant questions. Punjab experienced accelerated economic and population growth only after 1951. Economic growth was achieved essentially by agricultural development based on improved techniques of production, a process in which the state provided the basic rural infrastructure and individual producers responded vigorously to expanded economic opportunity (see Chapters II, III). Population increased because, while there was a steep decline in mortality, fertility fell more slowly. How far and in what ways were these economic and demographic variables interrelated? Some of the issues involved are discussed in the sections that follow.

2 Trends of Population Change

We have already observed in Chapter I that during the period 1906-46, the agricultural output in this region remained almost entirely stagnant. Changes in population during the first three decades of this century were also very slight. Between 1901 and 1931 the population of Punjab increased from around 12.17 to only about 12.57 million and part of this increase was simply the effect of improved coverage.

After 1931 the population increased at a higher rate
than before, and during the next two decades there
was an increase in numbers of about two millions. A
sharp acceleration in population growth came only
after independence, during the decade 1951-61, and
it has continued since.

Traditionally, since 1881 the proportion of the
male workforce in this region engaged in
agricultural activity had been around 61 per cent.
It increased between 1901 and 1931. In 1961, the
proportion of the male workforce in agriculture was
again 61 per cent; it rose to 65 per cent in 1971
and fell back to 61 per cent in 1981. Since the
absolute number of people dependent on agriculture
has continued to increase, the cultivable area per
agricultural worker has declined steadily,
especially since 1961. An extension of the area
under cultivation occurred during 1951-61, but this
process seems to have reached its natural limits, so
that future increases in output can for all
practical purposes come only from increases in
capital or improvements in technology.

The density of population has increased during
1951-81 along with a fast-growing agriculture and an
expanding infrastructure (Table 4.1). The effect of
these and the spread of knowledge about measures of
birth control should have hastened the transition
from a demographic phase with high birth and low
death rates to one where both are low. The speed
and duration of transition did not, however, follow
the historical pattern of Western Europe. We
examine in the next section what did happen.

3 Fertility, Mortility and Migration

Changes in population can result only from those in
fertility, mortality or migration. As we have
already remarked,from the point of view of the
population history of the region, the situations
before and after independence are qualitatively
quite different. Even though infant mortality had
started to decline around 1931, a sharp decline can
be observed only after 1951 and the decline has
continued, while the crude death rate appears to
have been reduced to almost half its previous level
in the course of a single decade (1961-1971).

Table 4.1 POPULATION GROWTH IN PUNJAB, 1901 to 1981

	Unit	1901	1921	1931	1951	1961	1971	1981
1. Population	Millions	12.17	11.41	12.57	14.83	18.73	23.59	29.52
2. Density of Population	Number of persons per square kilometer	129	121	133	157	198	249	312

Note: Punjab refers to Indian Punjab and Haryana states combined.
Source: Census of India, various reports.

Table 4.2 ADJUSTED BIRTH AND DEATH RATES IN PUNJAB 1961 to 1975*

Year	Death Rate (Adjusted)	Birth Rate (Adjusted)
1961	21.0	45.3
1962	21.0	44.9
1963	19.0	42.2
1964	18.6	41.0
1965	16.4	41.0
1966	16.4	38.4
1967	14.5	38.0
1968	13.8	36.0
1969	15.8	38.0
1970	14.5	37.1
1971	11.0	37.3
1972	11.9	36.0
1973	11.0	36.0
1974	11.9	35.3
1975	11.5	36.6

* For sources of data and methods of adjustment, see text.

69

Demographic Transition in Punjab

Before we proceed to discuss trends in
mortality and in fertility in the post independence
period, we shall make a few remarks about the basic
data on birth rates and death rates on which our
discussion is based. These are derived from Civil
Registration statistics published in the official
publication, Vital Statistics of India. That these
data suffer from serious inaccuracies, in particular
that for both birth rates and death rates they
considerably underestimate the true figures, is well
known. Indeed, because of their deficiences, some
scholars have held that this body of data cannot be
used for the purpose of research at all. If we were
concerned with forecasting the population of Punjab
in a future year, there would be undoubtedly much to
be said for such a view. However, that is not our
concern. Our purpose is simply to try to discover
whether mortality and fertility in Punjab in the
recent past show any discernable trend, whether they
have been rising or falling or have remained stable.
For this purpose, we believe in common with Davies
(1951) and Krishnamurty (1966) that registration
data, provided they are suitably adjusted, are quite
useful. It is these adjusted data, presented in
Table 4.2, which form the main basis of our
discussion below.
    For the later part of our period, we have
relied on data provided by the Sample Registration
Scheme, which are believed to be more reliable than
Registration statistics but are available only from
1968 (Table 4.3).
    Let us consider first the decline in fertility.
Birth rates seem to have been high and stable up to
1961 when a decline started. During the decade of
the 1960s the extent of the decline in the birth
rate was quite considerable. On what has happened
since, the picture is not quite so clear. The
figures for birth rate (adjusted) as given in column
3 of Table 4.2 appear to suggest that during the
first half of the 1970s the tendency of the birth
rate to decline more or less came to a halt,with
yearly fluctuations occurring around the same basic
level. On the other hand, if we turn to the figures
for the rural birth rate due to the Sample
Registration Scheme, and given separately for Punjab
and Haryana states (Table 4.3), a decline in
fertility seems to be continuing. The difference is

70

Table 4.3 RURAL BIRTH AND DEATH RATES IN PUNJAB AND HARYANA STATES, 1968 to 1978, SRS ESTIMATES

Year	Birth Rates (per thousand)		Death Rates (per thousand)	
	Punjab	Haryana	Punjab	Haryana
1968	33.6	n.a.	12.0	n.a.
1969	33.6	39.2	11.6	11.8
1970	34.7	38.0	11.8	10.0
1971	35.0	44.2	10.9	10.4
1972	35.8	42.2	13.4	12.3
1973	27.7*	n.a.	11.4	n.a.
1974	33.0	41.6	11.3	13.4
1975	32.5	39.9	11.3	13.2
1976	32.4	37.6	11.4	13.9
1977	31.8	35.7	11.4	14.9
1978	28.8	34.8	11.8	14.4

Notes: (i)   SRS denotes Sample Registration Scheme.

(ii)* Data is for the first half of the year only. For reasons of seasonality the birth rate reported during the first half is usually lower

Source: Sample Registration Bulletin, various issues.

perhaps more apparent than real, for it is in the years after 1975 that decline in the birth rate shown in the SRS figure is mostly concentrated. However, irrespective of which particular statistical series we may happen to be using, the rate of decline in fertility during the 1970s was clearly much slower than during the previous decade. Rural-urban differences in birth rates, death rates and child mortality are observable in Punjab as elsewhere. Urban rates in all these vital statistics are consistently lower. However, the rural-urban gap which was high in Punjab during the 1950s and 1960s has started declining, the decline being particularly noticeable after 1966; but a substantial gap in infant mortality still remains.

The decline of mortality, which started in the 1930s and accelerated after 1951, was due to some improvement in environmental hygiene (especially drinking water and sanitation), in health care facilities (see Table 4.4), nutrition and housing facilities (discussed in Chapter VI) and education (described later in this chapter). Of these, substantial improvement in environmental hygiene resulting from the state's investment in infrastructure and public health and higher standards of food and housing seem to have been the two most important components.[4] Because of the greater expansion of rural infrastructure as well as higher initial levels of mortality in rural areas, the decline in death rates, particularly after 1965, was also faster in rural than in urban areas. Thus, the popular view that the recent decline in mortality in less developed regions simply reflects advances in Western medical science and technology and is exogenous to the development process itself is quite untrue of Punjab, for investment in rural infrastructure, rise in output and consumption and fall in mortality were closely linked. Formal education of females in this region has remained rather low, and therefore could not have been an important factor in reducing infant mortality; but informal education which influenced attitudes of the people may have been quite important. There is, however, a disquieting feature of the figures on the death rate given in Table 4.2, which also calls for comment, viz. that from 1971 it seems to have ceased to fall. This impression of a relatively

Table 4.4 HEALTH INDICES IN PUNJAB 1960-61 TO 1974-75

	Per Capita Real Expenditure on Health and Medical Care *	Number of Doctors per Thousand Persons **	Number of Hospital Beds Per Thousand Persons **
1960-61	2.11	0.30	0.65
1974-75	14.99	0.64	0.64

Sources: * Yearly data on per capita expenditure on medical care and public
health (including Central as well as State Government expenditure) at
current prices are given in the Pocket Book of Health Statistics of
India, various numbers. These were deflated by the wholesale price
index (with 1960-61 as Base) to derive real per capita health
expenditure.
** Computed from data given in Health Statistics of India, various
issues.

Table 4.5 SOME INDICATORS OF TRANSPORT, COMMUNICATION AND TECHNOLOGY IN PUNJAB, 1960-61 to 1974-75

Year	Surfaced Roads per 100 sq. km.	Inhabitants Per Post Office	Postal Articles other than Money Orders handled per capita per year	Number of Radios per thousand Population	Circulation of Daily Newspapers per thousand Population	Yearly Electricity Consumption in kW.h per capita	Engine-Driven and Electrically operated Agricultural Pumpsets per thousand Rural Inhabitants
1960-61	1.49	4699	12.17	12.51	7.68	69.73	0.98
1974-75	38.88	2702	19.11	27.75	9.65	165.02	20.27

Source: Statistical Abstracts of Punjab and Haryana, various issues.

stable death rate having been established during the
1970s is confirmed by the SRS estimates of rural
mortality given in Table 4.3. Indeed, according to
these the rural death rate in Haryana seems to have
increased over the decade. Possibly this is a
temporary phenomenon and the effects of the
improvements described, which led to a decline in
mortality earlier, will re-assert themselves in the
long run. One cannot take this for granted,
however, for it is equally possible that the impact
of improvements in environmental hygiene, health
care, nutrition and housing on the level of
mortality may have spent itself by the mid-1970s and
that fresh efforts are needed to reduce the death
rate further. We shall return to this question at
the end of this chapter.

The explanation of the decline in fertility is
even more complex. Changes in age structure caused
by declining infant mortality and improved rates of
survival of female infants during the 1950s and
early 1960s would have increased the proportion of
females in the fertile age group (15 to 45) which
would tend to increase fertility. But the forces
tending to reduce it were stronger.

Although the marriage rate remained stable, the
increase in the age of marriage affected
nuptiality.[5] The proportion of married women,
especially among those in the age group 15 to 20,
declined. This was partly the effect of female
education, which kept girls in school longer, thus
postponing marriage. This would tend to lower
fertility though not perhaps by very much. Reduced
infant mortality probably would have increased the
duration of breast feeding which acts as a natural
birth control device. However, breast feeding
habits are socially determined and could not have
changed substantially during this period.

More important was modernisation which acts in
more complex ways to reduce fertility than
economists appear to have grasped. Reduced
dependence on 'chance factors' and, correspondingly,
an increased role of 'choice factors' in farm
production management could have created an
attitude, a sort of informal education, encouraging
farm families to plan their own reproduction also.
This is suggested, for example, by the fact that the
proportion of couples practising some sort of family

planning has been rising steadily in this region, without any major increase in effort on the part of the government in propagating family planning programmes.

Rising opportunity costs of time for females, particularly among small farmers and landless labourers, would have made child rearing notionally more expensive. This, coupled with a desire for more education and for improved nutrition and housing, would induce decision agents in the households to plan smaller families with improved human capital. This conforms to the notion of economic rationality of the household decision agents and in this sense is akin to the reasoning advanced in the New Household Economics. However, the demand theoretic formulation of the New Household Economics needs rather strong assumptions, for example perfectly competitive factor markets which may not apply in the Punjab situation (see Chapter V).

In Table 4.5 we present data on some selected economic, technological and social indicators which represent modernisation. Their individual role and impact are difficult to ascertain. Modelling their effect on fertility is a complex exercise and is beyond the scope of this volume. However, it seems to be a reasonable hypothesis that the trend towards modernisation, which the various indicators taken together clearly indicate, contributed to the decline in fertility.

We now turn to the role of migration. While Punjab, in addition to the shock of partition in 1947, has experienced substantial in-migration and out-migration, statistics on these are not systematically available. These suggest that most of the migration out of and into Punjab is economically induced and caused by pull factors.

Rural-urban migration in Punjab has been slightly lower than in the rest of India. The growth of urbanisation also has been slower. Smaller towns have grown as fast as larger cities in Punjab during 1951-81. For India as a whole, on the other hand, cities have grown faster than smaller towns.

Migration into some selected villages of Punjab as well as migration from these villages have been examined in detail by Singh (1979) and Singh and

Oberoi (1980). These studies suggest that
out-migrants from these villages are mostly skilled
educated workers while immigrants into these
villages have been larger in number than the number
of out-migrants, a finding which is at variance with
the historical experience of Western European
economic development. However, this particular
aspect of recent Punjab experience may turn out to
be a short-lived phenomena.

Our discussion suggests that migration has been
much less important than falling birth and death
rates in explaining either the rate of growth of
population or changes in its age-composition, which
we consider below.

4 Dependency Ratios and Household Composition

Population pyramids provide a convenient way of
representing changes in the age and sex composition
of a population over time.

Figures 4.1(a), (b) and (c) represent
population pyramids of Punjab for the years 1951,
1961, 1971 and 1981. These show an expanding base
for the age group 0-10 for 1951, 1961 and 1971 but
not for 1981, which confirms the evidence on
declining birth and death rates presented above.
One of the implications of this change is that the
supply of labour in Punjab will continue to grow at
a high rate for another 15-20 years even if the
birth rate itself is declining.

Changes in the age and sex composition of
households would tend to affect the
work-participation rate, i.e. the proportion of
total population that is in the workforce. For
example, an increase in the proportion of children,
due to a continued spell of high birth rates, would
ceteris paribus bring down the work-participation
rate.

Estimates of the work-participation rate for
males during 1951-1981 are presented in Table 4.6.
These show a reversal during 1971-81 of a previously
declining trend in work-participation, which is in
keeping with our observations above on successive
population pyramids during 1951-81. The reversal
occurred both in rural and urban areas.

Because of problems arising from changes in

definitions in different Censuses (discussed in
Chapter II), estimates of female work-participation
tend to fluctuate erratically especially for rural
areas; and the Census of 1971 has become notorious
for its underestimation of the number of female
workers. However, according to expert opinion (such
as Sinha, 1982) a comparison of numbers of female
workers given by the 1981 Census with those of the
1961 Census is less liable to error of this kind,
although the 1981 definition of a worker is a shade
more liberal, and it shows a small decline in the
female worker-participation rates in rural areas of
our region together with a substantial increase in
urban areas, during 1961-81. On balance, it
declined. These changes cannot be attributed to
changes in age-composition. However, the increase
in women's work-participation in urban areas is
probably the effect of an increase in their
employment opportunities in the service sector. In
rural areas, while a rise in the agricultural wage
rate would increase the opportunity cost of leisure,
increased household income would also induce a
greater demand by households for leisure,
particularly by female members, whose traditional
and heavy domestic duties are not usually recognised
as 'work'. In rural Punjab, the income effect may
have been stronger than the substitution effect,
leading to a decline in the female
work-participation rate; but the decline may also
be partly due to an increase in female education
(discussed subsequently in this chapter).

5 Children in the Labour Force

In traditional societies, the use of child labour in
economic activity within a household is seen as 'on
the job training' and the extent of its use depends
on the level of living. In subsistence-affluence
(Fisk,1965) the use of child labour will be minimal,
while under harsher situations it would be much
higher. In a capitalist mode of production use of
child labour for a wage is seen as an exploitative
activity. Economic reasoning would suggest that
child and female labour being a 'buffer' would be
used in peak periods of economic activity and at
other times only in activities with lower marginal

77

product. In Punjab we have a coexistence of both
modes of employment - within a household and for a
wage. Traditionally, less than 5 per cent of the
children in the age group 5-14 years work in
economic activity in Punjab; and they also
constitute about 5 per cent of the labour force.

Table 4.7 presents data on child workers in
Punjab for 1951-81. We notice that the proportion
of child workers in Punjab, which was relatively low
to start with, has slightly declined over time as
would be expected in a situation of expanding
elementary education and increasing economic
opportunities for adults. The decline is more
noticeable after 1961.

Increases in the wage rate would increase the
opportunity cost of not working and hence that of
full-time education; but would also have an 'income
effect'. Households owning land of a minimum
economic size would have a greater ability and,
possibly, because of their superior information, a
greater desire to reduce use of child labour within
the family and allow children to get schooling as
full-time students.

The net effects on landless labourers and other
poor households may be less certain. Increased
opportunity cost of labour due to a rising wage rate
would induce the landless labouring households to
increase children's participation in the labour
force, if, as seems likely, the income effect is not
sufficiently strong to offset it. Relevant evidence
on this point has been provided in the reports of
the Agricultural Labour Enquiry Committees of
1950-51, 1956-57 and 1964-65 and the rural labour
enquiry of 1974-75. These show that during the
entire period 1950-51 to 1974-75 the average daily
wage rate in Punjab has remained approximately half
of the adult male wage, and the extent of use of
child relative to adult labour has also remained
roughly at a constant level.

A slight decline in the children's
participation rate seems to have occurred among
cultivator households, while among landless
labourers it has remained virtually unchanged. This
is a consequence of a number of factors operating in
opposite directions. However, given the segmented
nature of the labour market in rural Punjab and the
importance of cultural factors which could affect

Demographic Transition in Punjab

FIGURE 4.1   POPULATION PYRAMIDS FOR PUNJAB , 1951 TO 1971

1951

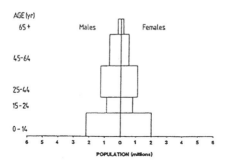

1961

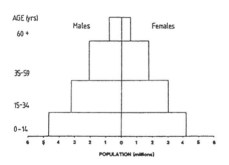

1971

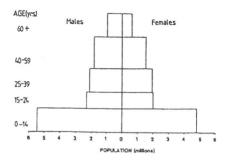

SOURCE: <u>Census of India</u> , various reports.

the use of child labour, it may not be particularly
useful to analyse the question of child labour in
terms of a systematic economic model.

6 Educational Characteristics and Quality of the
Labour Force

The proportion of literate males in the age group
15-35 is much higher in 1981 than in 1951; but for
females, the increase in literacy is not
particularly high and female education and literacy
rates remain much lower than those for males.
However, the analyses by AERC (1970) and Naik (1974)
bring out that the factors affecting participation
in education for males and females are essentially
the same, though the females suffer from a greater
handicap.
        School enrolment ratios for the age group 6-11
have improved spectacularly during 1951-81 for rural
as well as urban areas and also for females. For
the age group 11-14, improvement is much less
marked. As noted above, opportunity cost of child
labour particularly for the age group 11-14 seems to
be affecting it.
        In Table 4.8, we report the percentage of
persons of age 15-59 returning themselves as
full-time students in Punjab in 1961, 1971 and 1977.
The corresponding figures at an all-India level are
also given for comparison. For males there is a
noticeable increase in this percentage during
1961-71 in both rural and urban areas of Punjab,
while during 1971-77 there has been a small decline.
This could be interpreted as showing that the
opening up of better economic opportunities to the
uneducated together with an increasing risk of
educated unemployment have increased the opportunity
cost of full-time education for people aged 15 and
above and so made higher education less attractive
from an economic point of view. For adult females
the situation is in direct contrast with their
counterparts in lower age groups (for elementary
education) and also with males. The proportion of
females in the age group 15-59 reported to be
full-time students has been steadily rising since
1961, although it still remains extremely low.
Rural urban differences as expected persist but seem

Table 4.6 WORK PARTICIPATION RATES (Age 15 - 59)

Year	All India (Males)			All India (Females)		
	Rural	Urban	Total	Rural	Urban	Total
1961	93.78	83.20	91.60	50.76	18.32	45.17
1971	89.46	78.24	86.89	21.73	11.01	19.60
1977	87.67	78.31	-	39.71	19.13	-
Punjab						
1961	90.80	83.74		27.85	8.11	23.85
1971	87.93	80.26		2.35	4.84	2.91
1977	84.51	82.45		25.57	15.64	-

Source: 1961, 1971: Census, and 32nd Round, National Sample Survey.

to have declined during the decade of 1970s.

In Figure 4.2 we report the proportion of cultivators who have completed secondary education. The proportion of educated farmers has been steadily rising since 1961 and seems to have been a contributory factor in agricultural modernisation. Chaudhri (1979) analyses this process in detail.

In Table 4.5 we present inter alia some indicators of the development of communications facilities in Punjab during 1960-61 to 1974-75. These figures show that the circulation of newspapers increased relatively slowly, which reflects the slow progress of literacy among adults. On the other hand there was a spectacular increase in road mileage and substantial improvements occurred in such respects as access to and use of the post office, and the ownership of radios.

These indicators of improved communication when seen in conjunction with improved education indicate an accelerating trend towards modernisation and human capital accumulation. The enabling conditions were created by agricultural growth noted in Chapter II and the state's effort in creation of infrastructure discussed in Chapter III. The improvements in health and formal and informal education must have been unequally distributed among different sections of the population, but almost all sections of Punjab society seem to have derived some benefit from them (see Chapter VI), and this is reflected in the demographic trends discussed in this chapter.

Nevertheless, we must end this chapter on a cautionary note. Economic and demographic change are indeed, as we have argued, strongly inter-linked. Yet, it would be wrong to imply that one is simply determined by the other, as some of the more mechanical applications of the theory of demographic transition seem to suggest. Punjab experience amply bears out that the relationship is not automatic. Substantial economic progress was achieved in Punjab during the first two decades of independence, and a significant decline in mortality also occurred during the 1960s. During the 1970s, as we have seen, agricultural output continued to increase but the decline in mortality came to a halt. In order to make the people of Punjab healthier and more long-lived, continued economic

Table 4.7 DISTRIBUTION OF CHILD WORKERS IN PUNJAB, 1971
(Child Workers in '000)

	Male		Female			% of Child Workers to Total Workers	% of child Workers to Total Child Population	% of Scheduled Castes in Population
	Rural	Urban	Rural	Urban	total			
Haryana State	116	8	13	1	138	5.2	2.9	19
Punjab State	209	21	2	1	233	5.9	4.2	25
India	7277	607	2686	168	10,738	5.9	4.7	15

Source: UNICEF, An Analysis of the Situation of Children in India, UNICEF Regional Office, New Delhi, 1981.

83

progress is still essential, but it is not enough.
What is required now is a massive and concerted
effort to improve sanitation, making drinking water
safe, and to provide a simple yet effective primary
health care system in every village, linked to
hospital facilities in the smaller towns, facilities
which themselves must be expanded and improved.
Special attention is also required to the medical
care and health needs of females who, in Punjab as
elsewhere in India, have a lower expectation of life
than males.

This brings us to the important point that in
making policy decisions in this area, demographic
change must be considered as a whole: no policy
which does not fully recognise the linkage between
its fertility and mortality aspects could succeed.
In the context of Punjab, this means, for example,
making family planning services part of a package
which includes maternity and child care services as
well as elementary education for adult females.

It should be clear from what we have just said
that we are not advocating a policy of laissez faire
towards population growth in Punjab. The theory of
demographic transition, properly interpreted,
implies no such thing, although it has often been
supposed to do so. Indeed our discussion in this
chapter of recent trends in fertility suggests that
there is an urgent need for a policy to accelerate
fertility decline in Punjab. An essential condition
for such a policy to succeed is that it must be
integrated not only with policies for reducing
mortality but also with those for economic
development in general. This, in our view, is the
real point of a demographic transition approach, not
the economic determinism of some of its exponents.

Our argument has wide implications. It is
often said that the less developed countries of
today face much greater obstacles to economic
development than their counterparts in Western
countries in the past. A less favourable man-land
ratio and faster population growth are usually
mentioned as examples. The offsetting advantages of
a late start, which we may call the late development

Table 4.8   EDUCATIONAL INDICATORS IN PUNJAB, 1961 to 1975

| States | Social Indices Literacy Effective Exluding 0 – 4 Age | | | | | | No. of Literate Per 1000 Persons | No. of Female Literate Per 1000 | Female Secondary School Enrolment | |
| | General | | | Females | | | | | | |
	Total	Rural	Urban	Total	Rural	Urban			F – 1 (*)	F – 2 (@)
1961										
Punjab	29.18	-	-	17.07	-	-	242	141	12.6	4.7
1971										
Haryana State	31.91	25.92	58.89	17.77	11.10	48.14	269	149	21.9	9.9
Punjab State	38.69	32.08	59.97	29.91	22.99	52.13	337	259	35.5	15.0
1973										
Haryana State	31.91	25.92	58.89	17.77	11.10	48.14	269	149	21.8	10.5
Punjab State	38.69	32.08	59.97	29.91	22.99	52.13	337	259	33.7	15.6
1974 – 75										
Haryana State	31.91	25.92	58.89	17.77	11.10	48.14	269	149	23.0	11.4
Punjab State	38.69	32.08	59.97	29.91	22.99	52.13	337	259	34.9	16.5

Source: Statistical Abstract of India, various issues.

effect, have received less attention. However, these include the technology and know-how that have been built up in economically advanced countries and which could, in principle, be adapted to the particular circumstances of individual developing countries or regions. It is not just the medical technology that is relevant here, for spill-over effects on demographic variables may be caused by technical change in other areas as well. The late development effect is already noticeable in Punjab and it offers both challenge and opportunity for the future. How far such opportunity would be realised in practice depends on the nature of society's response to it.

NOTES

1. For a concise and informative summary of the debate on population and development, see Teitelbaum (1974).
2. As against Malthus's principle of population, Karl Marx pointed out that an abstract law of population existed in plants and animals alone and only in so far as man had not interfered with them. Marx, on the contrary, held that laws of population were specific to the mode of production prevailing in the society concerned and were historically valid only within its limits (Capital, Vol.I, p.62 ff). Marx's emphasis on socio-economic structure as determining the nature of population growth is more in line with recent work on economic-demographic interaction models than most of classical political economy which took a Malthusian stance.
3. In a sense, the development of high-yielding varieties could itself be regarded as an international response to population pressure on land. On the other hand, it is difficult to accept Boserup's argument that economies of scale make it relatively easier to provide rural infrastructure in densely populated countries, for the provision of infrastructure may depend more on policies followed by the state than on population density as such. Punjab, for example, which is now well supplied with infrastructural facilities and which, as Boserup herself has noted, is far from being typical of

Table 4.9 SOCIAL INDICES IN PUNJAB 1961 to 1973

| States | Females | | Female Workers (NA) | | Urban Population As % of Total Population | Density of Population per sq. km. | Life Expectancy at Birth (by Quasistable Method) |
	Average Age at Marriage	Workers As % of Female Population	As % of Female Population	As % of Total			
1961							
Punjab	17.59	14.20	2.79	35.1	20.1	166	47.5
1971							
Haryana State	N.A	2.41	0.86	33.12	17.66	227	50.6
Punjab State	18.8	1.18	0.97	36.37	23.73	269	43.83
1973							
Haryana State	N.A	2.41	0.86	33.12	17.66	227	50.6
Punjab state	18.8	1.18	0.97	36.37	23.73	269	43.83

Source: Statistical Abstracts of Punjab and Haryana States, various issues.

FIGURE 4.2   INCIDENCE OF EDUCATION AMONG FARM WORKERS : ORDERING ACCORDING TO INTER-DISTRICT
AVERAGES

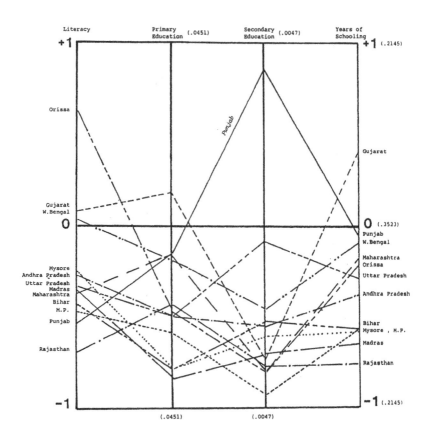

India, is no more densely populated than other regions of India (cf. Boserup, 1981, pp.203-4).

4. The question which of the two was more important does not admit a clear-cut answer. While it is true that improvements in food and shelter were not large enough to account for the kind of decline in mortality that occurred, it is equally true that such a decline cannot be maintained long without also maintaining nutrition and other necessary consumption at a reasonably high level. On this point see Eversley (1965, p.39) and Dasgupta (1980, pp.41-3).

5. Traditionally the age at the marriage ceremony itself was lower by several years than the age at which cohabitation started, an occasion which was normally marked by a second ceremony called gauna or muklawa. The age at marriage for village women in Punjab had been rising from about the mid-1920s but the age at start of cohabitation remained stable and, as a result, the customary delay between marriage and cohabitation disappeared. From the mid-1950s, however, the rise in the age at marriage was accompanied by a rise in the age at cohabitation as well. See John B. Wyoh et al., 'Delayed Marriage and Prospects for Fewer Births in Punjab Villages', Demography, Vol.3, No.1, 1966.

Chapter V

PRODUCT AND FACTOR MARKETS IN RURAL PUNJAB

Sowing seed in one's own land and reaping the harvest for distant markets are apparently contradictory. The seed-sowers naturally cling to the soil which they cultivate while the distributors of the harvest develop a different mentality.

Rabindranath Tagore

Market efficiency is considered desirable, and according to standard economic doctrine it will prevail if there is perfect competition. Perfect competition, the text books tell us, requires the existence of (i) a large number of buyers and sellers functioning without collusion or agreement; (ii) perfect knowledge about market conditions plus the logical utilisation of such knowledge by all participants in the market; (iii) homogeneity of products; and (iv) free mobility of buyers, sellers and products. Reality rarely conforms to this text book definition. Tests of how efficiently markets are actually working are usually based on some notion of divergence from an 'ideal type' of perfect competition, giving rise to considerable scope for disagreement. There have been a number of studies of this kind done in Punjab, but they have reached different conclusions. We shall discuss the prevailing conditions and some of these studies in this chapter. The chapter is divided into five sections. The first deals with the product market and in the second we examine the question of factor proportions and scale economies. The next two deal with labour and capital markets respectively. The last attempts an overall evaluation.

1 Product Market and its Integration

The Royal Commission on Agriculture (1928) came to the judgement that markets for agricultural products in Punjab were neither competitive nor efficient.

Product and Factor Markets in Rural Punjab

According to the Commission, conditions prevailing
in land and capital markets were even worse and
certainly oligopolistic. The declared purpose of
much of government legislation since then has been
to minimise the more obvious inequities in the
functioning of the market, e.g. collusion between
buyers and middlemen against sellers (farmers) or
short changing in weights and measurements, and
manipulated price quotations at auctions in grain
markets (which have been used as a means of price
setting in this region for centuries). By 1953 the
regulation of markets, the establishment of
independent Umpires and the separation of
auctioneers (known as Kacha Arathiya) from the
buying agents (known as Pucca Arathiya) had
certainly brought about some improvement in market
functioning. The establishment of a market
intelligence service and the use of All India Radio
to broadcast grain prices in different markets of
Punjab must have gone a long way towards integrating
the market and improving its efficiency. On the
other hand, there were some restrictions on the
movement of grains across district boundaries in the
1950s while restrictions across state boundaries,
which were part of the Food Zones system, continued
till the 1970s.[1] In this respect the grain market
was certainly not perfectly competitive.

The working of the grain markets in Punjab has
been examined by numerous scholars, including
Cummings (1967), Jassuwala (1967), Mellor (1968) and
Lele (1971). Using similar procedures for measuring
marketing margins and the profitability of grain
trade, these researchers came to the general
conclusion that grain markets in Punjab were
functioning efficiently and competitively. Both
this conclusion and the methodological approach used
in reaching it came under attack from Sarkar (1979)
and more recently Rudra (1982). The latter, for
example, points out that Pareto optimality itself is
a very demanding condition and cannot be achieved
without the prevalence of competition in all factor
and product markets simultaneously. Nevertheless,
all things considered it seems reasonable to
conclude from the empirical evidence that markets
for agricultural products in Punjab have been
getting more integrated and are more efficient than
they were.

Product and Factor Markets in Rural Punjab

2 Factor Proportions and Scale Economies

Farm management studies carried out in various
regions of India, including Punjab during the mid
1950s, indicated that as compared to large farms
smaller farms were characterised both by higher crop
output per acre of land and by the use of greater
quantities of labour as well as other variable
inputs per acre of land. These findings were
analysed by Sen (1962) who pointed out that smaller
farms seemed to have been cultivated more
intensively than the large farms and they provided
greater opportunities of labour use even though, in
terms of profitability, their enterprise remained
unremunerative; large farms, by contrast, had a
positive profit. This conclusion and the related
question of the existence or otherwise of
diseconomies of scale in agricultural production
were examined by among others Mazumdar (1963), Saini
(1969,1979), Yotopoulos and Lau (1971), Anderson and
Dillon (1971) and Rao (1974), who used production
function estimates as well as other types of
econometric analysis for this purpose. A good
analytical summary of this literature can be found
in Berry and Cline (1979). Recent farm management
studies for the post Green Revolution period, i.e.
the late 1960s and early 1970s in the Punjab region,
suggest that factor proportions continue to be
different in different farm size groups. However,
small farms do not appear to be using more of <u>all</u>
variable inputs per unit of land. They continue to
use a larger amount of labour per acre as compared
to large farms; but large farms use a greater
quantity of chemical fertilisers per acre. The
inverse relationship between farm size and output
per acre that had been observed in the 1950s also
seems to have disappeared by the early 1970s,
according to some recent studies such as Heady
et al. (1973), Rao (1976), Kahlon and Singh (1977),
Bardhan (1972), Saini (1979) and Roy (1981) , most
of which used production function analysis to derive
their results.
    The main results of production function studies
using Punjab farm management data for different
years of the 1950s and the 1960s are given in Table
5.1. All the studies used the Cobb-Douglas
production function of the unrestricted variety

Table 5.1 SELECTED PARAMETERS OF COBB DOUGLAS PRODUCTION FUNCTIONS REPORTED IN DIFFERENT STUDIES ON PUNJAB FARM DATA

Study	Year	N	Land	Labour	Fertiliser	Sum of Elasticities	Return to Scale	Remarks
Saini (1973)	1955-56	200	.4468 *	.8242 *	.0108	1.06	Constant	Bullock labour exluded due to
	1956-57	200	.4322 *	.4149 *	.0151	1.01	"	multicollinearity
	1955-56	200	.6138 *	.7297 *	.0129	1.06	"	Including bullock labour
	1956-57	200	.6139 *	.3443 *	.0108	1.05	"	"
Small Farms	1955-56	42	.3798 *	.6579 *	.0534	0.86	"	Bullock labour exluded due to
Medium Farms	1955-56	69	.4617 *	.9793 *	-.0044	0.94	"	multicollinearity
Large Farms	1955-56	89	.4657 *	.8317 *	.0667	1.10	"	1955-56 S,M,L class Regression
Small Farms	1956-57	43	.8048 *	.1039	-.0371	1.00	"	co-eff. equal. 1956-57 S,M,L
Medium Farms	1956-57	61	.7070 *	.5109 *	.0029	1.50	Increasing	class co-eff. different at 5%
Large Farms	1956-57	96	.4232 *	.4569 *	.0394 *	1.01	Constant	level.
Bardhan (1973)	1955-56	98	1.0934 *	.3492			Increasing	All villages of 1955-56 included
Ferozepur	1955-56	80	1.1073 *	.3854			"	Common 8 villages of 1955-56 and
Amritsar	1955-56	100	1.0474 *	.1803			Constant	1967-68, bullock labour exluded from all regressions.
Srivasta, Heady, Nagadevare (1973)	1955-56	197	.3764 *			.96	Constant	
Krishna (1964)	1954-55		.31 *	.21 **		.87	Diminishing	
	1955-56		.55 *	.39 *		.97	Constant	
	1956-57		.18 *	.40 *		.95	"	
Bardhan (1973)	1967-68	150	1.0646 *	.1794 *	.0510 *			All villages of 1967-68 included.
Ferozepur	1967-68	80	1.0967 *	.3629 *	.0367 **			Common 8 villages of 1955-56 and
	1967-68	136	0.9895 *	.3034 *	.0217 *			1967-68 wheat farms only.
Srivasta, Heady, Nagadevara (1973)	1967-69	299	0.3928 *	.5097 *		1.19 ***	Increasing	Two years combined data with year Dummy introduced.
Chaudhri (1979)	1961-64	1038	0.236 *	.226 *	.132 *	1.03	Constant	Agr. Econ. Research Centre data Complete enumeration of 19 villages.

Notes: * Significant at 1% level, ** Significant at 5% level, and *** significant at 10% level.

which enabled them to test for returns to scale;
but the specification of the production function was
somewhat different in different cases. We have
reported in Table 5.1 the computed elasticity for
land, labour and fertiliser only, whereas in each
case the number of explanatory factors used was
greater than these three. In addition to reporting
the parameters of the Cobb-Douglas production
function relating to land, labour and chemical
fertiliser we have also reported the observed sum of
elasticities including the result of the statistical
test of the elasticity being significantly different
from unity. Unfortunately, the results are not such
as to provide much reason for confidence in their
reliability. Although in each of these studies, the
coefficient of land was found to be statistically
significant at the 1 per cent level of significance,
the magnitude of the land coefficient as estimated
by various authors varied from 0.18 to 0.804. Even
if we consider only the period of the mid-1950s, the
variation in the land coefficient estimated by
different studies is quite considerable. In the
case of studies using 1967-68 to 1969-70 data, the
variation of the estimate of the land coefficient is
even greater. A similar problem arises for the
coefficient of labour, viz. its estimates are very
different in different studies, even though the
estimate is found to be statistically significant in
each case. As regards their overall conclusions
about the nature of returns to scale the studies do
not always agree. If one accepts the specification
to be complete in all these studies and the
estimating procedures to be valid, then according to
the reported statistical results of Srivastra,
Nagadevara and Heady (1973) returns to scale in
Punjab agriculture for this set of data are
increasing for 1967-68 and 1968-69; while Bardhan
(1973), using 1967-68 data only, reports constant
returns to scale. In view of instability (across
studies) of the reported coefficients of the
Cobb-Douglas production function, it is difficult to
conclude that these studies have succeeded in
estimating the true elasticities of input
substitution for Punjab agriculture. Their failure
to do so could be due to wrong or incomplete
specification or to some inherent defects of the
production function approach itself. One possible

source of mis-specification is that farms of
different size groups may also differ in respect of
the kinds of technology they mainly use, and the
possibilities of factor-substitution may depend on
the technology used.

Some parameters of Cobb-Douglas agricultural
production functions for Punjab, based on farm
management data for Ferozepur district (1968-69,
1969-70) for some sub-groups of farms using
different technologies of production, are reported
in Table 5.2. While the statistical results are
significant in four out of six cases, the
coefficients are very different as between
sub-groups of farms using different technologies of
production. The value of the land coefficient
estimated for 36 farms using 'canal irrigation and
bullock-power' is 1.04; while for a group of 64
farms using 'tractor and tubewell' technology, the
value of this coefficient is estimated at 0.3543.
Differences in the technology of production could be
quite important in the context of our present
discussion but such differences and their
implications for input substitution have been
neglected by studies using a production function
approach. There are also more fundamental problems
involved in the use of a production function
approach, viz. those related to the measurement of
capital, considered in Chapter II. For these
reasons, we would not place too much reliance on
estimated coefficients of a production function.
However, two generalisations can be made. First,
the specific advantage of small farmers in terms of
scale diseconomies observed in the situation of the
1950s is no longer effective. Returns to scale, in
all probability, are now either constant or
increasing. If they are not already increasing, the
penetration of capital into agriculture in the form
of farm machinery and equipment, the use of chemical
fertilisers and the like, would have a tendency to
negate scale neutrality and would indicate greater
economic advantages accruing to larger sized farms
in future. This is an important symptom of
transition in agricultural development.[2] Secondly,
factor proportions used on different farm size
groups are certainly different. Large farms appear
to be substituting labour for other inputs, for
chemical fertilisers in particular.

Table 5.2 SOME PARAMETERS OF THE COBB DOUGLAS PRODUCTION FUNCTIONS FOR PUNJAB
FM DATA 1968-69 TO 1969-70

Year	N	Technology	Land	Labour	Fert.	Sum of Elasticities	RS	$\bar{R}^2$
1968-69	46	Bullock-Canal	1.0367*	0.1912	-0.0722	1.0365	Constant	0.8953
	24	Bullock-Tubewell	-0.0937	0.3542	0.1699*	0.8419	Constant	0.8397
	14	Tractor-Canal	0.2962**	0.0663*	0.2671*	1.2997	Constant	0.9440
1969-70	64	Tractor-Tubewell	0.3543**	0.3536*	0.2443*	0.9472	Constant	0.8571
1968-69	150	All Farms Pooled	0.0539	0.4146*	--	0.5748	Diminishing	0.3129
1969-70	150	All Farms Pooled	0.4846	0.4301*	0.0954*	1.0203	Constant	0.8943

Notes: * Statistically significant at 1% level. ** Statistically significant at 5% level.
Source: Reported in Khalon and Miglani (1972).

3 Labour Markets and Modes of Employment

There could be many different explanations of factor
proportions being different as between small and
large farms. In the specific context of Punjab, an
explanation that has received particular attention
from scholars is in terms of different market
conditions being faced by farms of different size
groups. For explaining higher labour input on
smaller farms, it has been suggested (e.g. Sen,
1962, 1964; Mazumdar, 1965) that the opportunity
cost of family labour to farming household may be
less than the market wage rate. Bharadwaj (1974),
while arguing forcefully about the imperfections in
the agricultural labour market and basing her
arguments on FM data for 1950s, states 'Except
Punjab (where outside employment is insignificant)
off- farm employment tapers off as the size of
holding rises' (p.A-16). The question is why
off-farm employment is uncommon and rather
insignificant among small farmers in Punjab. One
important reason could be that, in agriculture,
working on the farm and working off the farm are not
purely additive activities. A given total quantity
of labour input in standard days in a year could
conceal considerable variations due to seasonality
of labour requirements, but the timing and
sequencing of labour inputs often have an important
effect on the level of agricultural productivity.[3]
It is possible that small farmers are able to take
wage employment only in the lean period when it is
not easily available. It is also possible that they
are reluctant to hire out their own labour for
reasons of prestige. This may be an important
reason in the case of females and children who may
be a resource specific to the family farm, but may
not be entrants into the wage labour market. Table
5.3 indicates that annual labour put in by a family
farm worker is, in most cases, lower on small farms,
and is highest on the largest farm size group. In
spite of this lower number of standardised days
worked per year, small farms also hire in labour -
not only casual labour but permanent labour also.
This is partly because casual labour at peak periods
is not easily available and is higher priced in peak
periods, and also because some agricultural
operations require at least two persons to be

working simultaneously. The possibility that for reasons of caste or status farmers may be unwilling to perform certain manual agricultural operations themselves does not appear to have such relevance in the Punjab context, as far as work done by males is concerned. Another explanation of the greater use of labour on small farms is provided by the working

Table 5.3 COMPARATIVE ANNUAL INPUT OF A FAMILY WORKER AND FAMILY FARM SERVANT - PUNJAB

Size Group	Days of 8 Hours	
	Family Farm Worker	Permanent Servant
Punjab 1954-57		
Below 5 acres	224	204
5 - 10 acres	248	231
10 - 20 acres	280	306
20 - 50 acres	295	258
50 and above	389	375
Average	279	325
Ferozepur 1968-70		
Below 14.82 acres	288	372
14.82 - 22.24 acres	307	292
22.24 - 34.59 acres	309	301
34.59 - 59.30 acres	328	333
59.30 and above	325	312
Average	298	323
Punjab (All Regions) 1967-70		
Below 10.00 acres	273	309
10.00 - < 17.50 acres	292	305
17.50 - < 25.00 acres	309	376
25.00 and above	319	330
Average	300	342

Source: Farm Management Studies Reports, different years.

of the factor-substitution principle under competitive conditions. If land is taken as given or 'fixed' for each holding (a very realistic assumption under conditions of Punjab agriculture), and if all producers face the same input and output prices and are on the same production function which allows for factor substitution, small farmers will choose a higher labour-land combination with higher productivity per acre and lower productivity per man-day. Thus as a counterpart to the inverse relation between the productivity of labour and the size of holding which is precisely what is observed in Punjab situation, for 1954-57 but not for 1967-70, Krishna (1964) reported results relating to 1956-57, and we have computed these for 1967-70, which are given in Table 5.4.

That the imperfections of the land rental market might significantly influence the input-output relationship on the different types of farms has been suggested inter alia by Bhagwati and Chakravarty (1969), Bhaduri (1973) and Bharadwaj (1974). It has also been emphasised in this context that if the land-owning class in the village is also the money-lending class the tenants could be at a serious disadvantage, for in such a situation tenants have to lease in land and borrow not only the working capital but also their own means of subsistence from the same source. In these circumstances, factor-proportions as well as the output per unit of land could vary as between tenant and owner farmers. This situation is not so relevant for Punjab, and the explanation for differences in factor proportions probably lies elsewhere.

4 Capital Market, Rural Credit and Farm Size

It is possible that access to the capital market is not equally easy for farmers belonging to different size classes of holdings. We have computed the amount of borrowing by farmers in different size classes of farm holdings according to the amount borrowed, the source of finance and the purpose for which it was utilised from farm management studies for all regions in Punjab and also for the years 1967-68 to 1969-70. These statistics are reported

Table 5.4 COEFFICIENT OF SIMPLE CORRELATION BETWEEN AVERAGE PRODUCTIVITY TO FARM SIZE
PUNJAB : LINEAR EQUATIONS

Productivity or Cost Variable (Dependent)	Size Variable (Independent)	Krishna (1964) 1956-57			Our Estimates @ 1967-70		
		Linear Regression Coefficient	Linear Coefficient Determination	Quadratic Coef. of Determination	Linear Regression Coefficient	Linear Coef. of Determination	Quadratic Coef. of Determination
Output per Acre (Y/L)	Acreage (L)	-.40 * (.14)	.038	-	-.1013 (4.844)	.040	-
Output per Manday (Y/M)	Acreage (L)	.034 * (.008)	.078	-	1.270 (1.1884)	.026	-
Average Cost C(Ca)	Acreage (L)	-.0013 * (.0004)	.003	.002	-.0028 * (.0008)	.318	-
Average Cost C(CC)	Acreage (L)	-.00133 (.00072)	.017	-	-.0096 * (.0012)	.700	-
Output per Acre (Y/L)	Output (Y)	-.0056 * (.0010)	.075	.106	.0015 (.0024)	.023	.131
Output per Manday (Y/M)	Output (Y)	-.00014 * (.00006)	.022	.226	.0007 (.009)	.019	.026
Average Cost A (Ca)	Output (Y)	-.0000089 * (.00000032)	.036	.045	-.0000016 * (.00000038)	.383	.432
Average Cost C (Cc)	Output (Y)	-.000028 * (.000005)	.130	.146	-.0000050 * (.00000059)	.731	.745

Notes: @ Our estimates are based on farm size group averages for 3 years 1967-68, 1968-69 and 1969-70 for the whole of Punjab and for Ferozepur District, giving us in all 27 observations over three years.

Figures in brackets are standard errors.

* Significant at 1% level.

in Table 5.5. The extent of indebtedness for a unit
of land holding is roughly the same among farmers of
different farm size groups but there seems to be a
considerable difference in their 'sources' of
borrowing. We find that for Punjab as a whole,
farmers in the highest size group obtained on
average nearly 90 per cent of their credit from
government and cooperative agencies whereas farmers
in the smallest size group obtained about 70 per
cent of their credit from these sources. The
traditional source, namely the village money
lenders, is insignificant for the highest farm size
group while it still contributes about 20 per cent
of all borrowings by small farmers. We do not have
information on the actual rate of interest paid by
farmers in each size group, but we do know that the
rate of interest charged by the state and
cooperative sources varied between 7 and 9 per cent
while the rate of interest charged by the money
lenders varied between 18 and 100 per cent. In all
probability small farmers are able to obtain credit
from state and cooperative sources for their
agricultural production needs but have to depend on
the money lenders for consumption credit. We have
no information about the credit needs of farmers in
different farm size groups nor do we know the extent
to which these needs are met by state and
cooperative sources. In the absence of this
information it is difficult to make any informed
judgements about the existence of 'internal capital
rationing' among small farmers which might be
contributing to their lower input of fertiliser and
other inputs per unit of land during the late-1960s.
As regards fixed investments per acre, there are
also systematic differences between small and large
farms. For example, larger farms have greater
quantities of major implements and farm machinery
per acre, and a larger stock of milch cattle per
acre, while smaller farms have a larger stock of
other cattle and irrigation equipment per unit of
land. That the larger farms have been investing
more in improved implements is clear in Table 5.6
from the average value of traditional and improved
implements per hectare of land in 1967-70 calculated
in prices then prevailing in different size groups
of holdings. On balance we find that there are
significant imperfections of the input market, in

Table 5.5 SOURCES OF CREDIT AND ITS USE, PUNJAB FARM MANAGEMENT STUDIES, 1967 to 1970 (Average)

District Year	Size-Group (Acres)	Average Size (Acres)	No. of Holdings	Loan per Holding (Rs)	Loan per Acre from State and Coopera-tives (Rs)	Source of Credit-Percent			Utilisation - Percent			
						States and Coopera-tives	Money Lenders	Others	Purchase of land	Input	Invest-ment	Non Pro-ductive
Punjab-All Region 1967-70	<10	6.73	55	336.84	35.35	70.22	20.82	8.96	-	37.40	42.04	20.56
	10-<17.5	13.43	63	630.28	33.51	72.56	8.81	18.63	2.17	45.37	45.90	6.56
	17.5-<25	20.81	65	772.26	33.69	91.07	3.98	4.95	1.58	52.02	35.39	11.01
	25 +	35.11	87	1244.32	31.68	89.07	3.08	7.85	4.35	58.62	27.14	9.89
	Average	20.87	270	803.48	33.19	85.76	5.42	8.82	2.55	53.44	30.50	13.51
Ferozepor 1967-70	<14.83	10.72	33.7	1114.54	56.68	54.52	38.03	7.45	N.A.	N.A.	N.A.	N.A.
	14.83-22.24	18.58	29.3	1989.96	88.43	82.57	10.97	6.46	N.A.	N.A.	N.A.	N.A.
	22.24-34.59	27.75	36.7	1424.33	31.41	80.92	14.49	4.59	N.A.	N.A.	N.A.	N.A.
	34.59-54.30	43.24	37.7	3501.94	54.59	67.40	27.93	4.67	N.A.	N.A.	N.A.	N.A.
	54.30 +	83.72	12.6	3979.52	44.09	92.76	1.26	5.98	N.A.	N.A.	N.A.	N.A.
	Average	30.74	150.0	2202.02	51.36	71.71	22.80	5.49	N.A.	N.A.	N.A.	N.A.

Source: Farm Management Studies Report, 1967-70.

particular the credit market in rural Punjab, which
must be partly responsible for the observed
differences in factor-proportions.

5 Fragmentation and Integration of Markets

Farms belonging to different size groups in Punjab
differ in respect of their use both of family and
hired labour inputs. This can be attributed to the
segmented nature of the rural labour market. Some
segments, such as household female and child labour
among small and middle farmers are not an integral
part of the labour market at all. The coexistence
of different modes of employment has implications
for the level of labour use and the choice of
capital intensity; these implications have been
widely discussed in the literature (cf. Sen, 1975),
and so has the impact of the introduction of modern
technology in agriculture on the nature of the
labour supply (see Heller, 1976). In all
probability, the introduction of tubewell irrigation
and high yielding varieties of seeds together with
the accompanying package of inputs and agricultural
practices increased the demand for labour used
during the slack season for weeding and other crop
husbandry operations. Substantially increased peaks

Table 5.6 INVESTMENT IN TRADITIONAL AND IMPROVED
IMPLEMENTS PER HECTARE- FEROZEPUR
1967-1970 (Rs.)

Holding Size Group (Hectare)	Implements Per Hectare	
	Improved	Traditional
Below 6	119.39	55.13
6 - 9	193.00	47.22
9 - 14	266.28	53.95
14 - 24	414.05	41.86
24 and above	409.85	16.53
Average	330.60	40.20

Source: Chaudhri (1979).

of labour demand were, however, chopped off by the introduction of tractors for ploughing and sowing and the partial mechanisation of harvesting operations. Thus agricultural mechanisation in the region has led to a smoothing out of peaks and troughs in the demand for labour. Different segments of the labour market have adjusted their supply differently in response to changed demand conditions.

Because of the rising opportunity cost of labour, the increasingly perennial nature of employment and the fact that farm work, thanks to mechanisation, has become in many respects less physically onerous and more technical in nature, the input of family labour has increased among medium and large farmers. The employment of permanent wage-labour appears also to have increased somewhat but the demand for seasonal casual labour has declined. As a result some displacement of casual workers from agriculture seems to have occurred. Since there has been relatively little industrial development in this region, this group of workers would tend to face an increasingly difficult situation. Some possible implications of such a development for income distribution are touched on in the next chapter and its contribution to the process of differentiation among the peasantry in Chapter VII.

The capital market, as noted above, has been expanding and is also probably tending to be less imperfect. However, large farmers have a much greater access to a cheap, stable and expanding supply of short term and long term credit from institutional sources. Small farmers on the other hand still have to rely heavily on a costlier and more erratic source of credit, namely money lenders. The problem of internal capital rationing among small farmers (noted in Chaudhri, 1968) seems to have largely disappeared with the coming of the new technology in agriculture since all groups of farmers are now using (otherwise scale neutral) modern inputs and new technology. The semi-feudal model of the landlord acting simultaneously as a monopsonist in the labour market and monopolist in the capital market does not appear to be applicable to Punjab.

NOTES

1.The Zonal System was changed a number of times during the last 25 years. For details of these changes see Kahlon and Tyagi(1982).

2. The emergence of economies of scale is usually one of the consequences of penetration of capital in the agricultural sector. This question has been discussed by many writers, for example Chaudhri(1974).

3. Timing of labour input is crucially important in agricultural operations. For technical aspects see ICAR(1974).

Chapter VI

INCOME DISTRIBUTION, POVERTY AND LEVELS OF LIVING

Of the rich, the main dish is meat; of the middling, it is milk; of the poor it is salt.

Mahābhārata

...the manner in which wealth is distributed in any society depends on the statutes and usages therein prevalent.

John Stuart Mill

For nearly two decades real agricultural output in Punjab both absolutely and per head of the population has grown at rates that are extraordinarily high. Economic growth, however, is not valued for its own sake. Its justification lies ultimately in the increased levels of economic welfare that it may help to achieve. How far has rapid growth in Punjab succeeded in bringing about higher levels of living for the people of the region? In particular, how far has it helped to bring about a decline either in the absolute incidence of poverty or in relative economic inequality? These are the questions that we propose to examine in this chapter. It is organised as follows. We begin with a general analysis of the relationship between inequality and growth in the process of agricultural development. The second section deals briefly with trends in average levels of consumption in Punjab during this period. The next two sections discuss trends in inequality and poverty respectively. The final section provides a summing up.

1 Growth and Inequality in the Process of Development

The first question that we shall discuss is the relationship between inequality and growth. In the context of agricultural development what kind of relationship, if any, should we expect between them? The received doctrine on the relationship

between economic development and the extent of
inequality was first stated by Kuznets and is
commonly described as the inverted U hypothesis,
i.e. as an economy grows from an initially low
level of development, inequality first increases,
reaches a peak, and subsequently declines. It is
important to distinguish between two aspects of the
postulated relationship. The first consists in the
observations that in inter-country comparisons, less
developed countries show greater inequality than
more developed countries[1]; that within the less
developed countries themselves, those at the very
bottom in terms of per capita income show lower
inequality than others, and more generally, that
countries at an intermediate state of economic
development show a more unequal distribution of
income than the richer or poorer countries (Kuznets,
1955, 1966; Adelman and Morris, 1973; Paukert,
1973; Chenery and Syrquin, 1975; Ahluwalia, 1974
and 1976; Lydall, 1977; Fields, 1981).
The second aspect of the Kuznets hypothesis
relates to causality. Observed correlations between
development and inequality, however interesting in
themselves, cannot be of general relevance unless
the nature of a causal mechanism at work can be
established. Kuznets' own argument turns primarily
on the changing structure of the economy as
development proceeds. Economic development, in his
view, is brought about essentially by a shift of
labour from a traditional, low-productivity,
low-income sector (agriculture) to a modern,
high-productivity high-income sector (industries and
services). Since such a process necessarily
benefits asset owners (capitalists) in the modern
sector and the modern sector workforce but leaves
the real income levels of those remaining in the
traditional sector much the same as before, an
increase in inequality is seen as an inseparable
part of the growth process itself. Higher rates of
saving and asset accumulation in the modern sector
(profits being almost entirely reinvested while
wages are consumed) support both the accelerated
growth of output and an increase in inequality.
The falling part of the inverted U-curve
relating inequality to aggregate income is harder to
explain in purely economic terms; but growth in
numbers of workers in the modern sector could itself

encourage the formation of trade unions, which not
only strengthen their bargaining position vis-a-vis
their employers but help them for the first time to
gain some influence on matters of public policy
affecting income distribution. How far inequality
would be reduced, in consequence, would vary with
prevailing social and political circumstances,
however. Clearly, the logic of the Kuznets
hypothesis relates to changes occurring over time;
but the empirical studies which have helped to make
it generally accepted are based on regression
analyses of inter-country cross section data. Few
have directly examined changes in inequality over
time in a particular country or region; the lack of
adequate time-series data on income distribution
makes this difficult.

A notable exception is Soltow's (1960) attempt
to explore the impact of the Industrial Revolution
on income inequality in Britain. If, he argues, it
was true that a dynamic industrial group developed
which made the rich richer, even if the poor did not
become poorer, relative income inequality among all
groups would increase, as indeed the Kuznets model
suggests. By estimating Lorenz Curves of personal
income distribution in Great Britain in selected
years, he found that inequality did not change
during the eighteenth and nineteenth centuries, but
fell sharply after the First World War. He explains
the failure of inequality to rise in the earlier
period by the fact that in pre-industrial Britain, a
substantial proportion of total income came from
landed property which was very unequally
distributed, and that this proportion diminished
with industrialisation. His argument that the
Industrial Revolution could not have introduced an
element of greater inequality than property income
from land in a traditional society has a wider
analytical relevance, pointing in a different
direction from Kuznets.[2]

Neither of these hypotheses is directly
applicable to the type of development with which we
are concerned in this study, in which the impetus to
growth came from the agricultural sector itself, and
the proportion of the labour force in agriculture
showed relatively little decline. Formally, one
could still identify modern and traditional sectors
within agriculture, the former referring to the

larger farmers working with large quantities of
capital equipment, using modern technology and
producing high levels of value added per worker, and
the latter referring to smaller farmers using little
capital and traditional technology and generating
low levels of productivity. The Kuznets analysis
could then be applied to this dualistic agricultural
economy consisting of two sectors so defined. Our
discussion in the previous chapters suggests that at
least in the Punjab context, such an approach would
not be particularly helpful. A strict dichotomy of
Punjab agriculture into traditional and modern
sectors- small-scale farmers being identified with
the traditional sector and the large farmers with
the modern sector- is no longer tenable. It would
be more useful to analyse the process directly.
From this point of view we could examine two sets of
considerations affecting inequality of real incomes,
namely those related to inequality across farm size
groups and those relating to inequality between
landless agricultural labourers and others. As
regards the first of these, the crucial factors are
how fast new modern technology would be adopted by
farmers belonging to different size groups, and how
far the new technology can be regarded as
effectively scale neutral. The general view taken
in the literature has been that the HYV technology
is scale neutral in its production aspects although
this is partly offset by inequalities in access to
credit, inputs and the use of rural infrastructure,
all of which are considerably influenced by the
higher social status and economic power of larger
farmers. Because of these, and perhaps also because
of lower risk aversion, it was the larger farmers
who by and large first adopted the HYV technology in
the Punjab in the mid-1960s. However, smaller
farmers and indeed even the marginal farmers rapidly
followed.[3] Indeed the rate of adoption of high
yielding varieties, as that of many other
innovations, tends to follow a logistic pattern with
its characteristic S shaped curve in which smaller
farmers follow larger farmers in the use of a new
technology with a lag.[4] In the case of Punjab, the
lag appears to have been fairly small. It follows
that the extent of inequality accruing from income
differences between farmers of different size groups
may be expected to rise sharply at first and then

start declining.[5] This general picture may,
however, be complicated by certain other
considerations. First, even before the HYV was
introduced larger farmers in Punjab were often
already using improved local varieties. Increase in
income and productivity brought about by the
adoption of HYVs could, therefore, have had
relatively lower impact in their case as compared to
those who switched directly from traditional to
modern technology. If so, the effect of new
technology in increasing inequality would be reduced
to some extent.

Secondly, technology includes both biological
and mechanical components. Studies on adoption
referred to above are invariably based on the
former, i.e. they relate to the behaviour over time
of the proportion of crop area under High Yielding
Varieties. Hoeever the adoption of such implements
as tractors or threshers may follow a very different
pattern, for example one based on the income
threshold hypothesis discussed earlier in the
context of durable consumption goods (Pyatt, 1969;
Davies, 1974). To this extent, given that the
initial income distribution was highly skewed,
mechanisation would tend to make it even more so.
Thirdly, as the mechanical components just referred
to become increasingly important, the scale
neutrality of the new technology which we have so
far assumed may no longer hold, and indivisibilities
may lead to small farm size groups being at a cost
disadvantage.

A more general question, namely how far can
economies of scale exist in agricultural production,
is of some importance here. It has, for example,
been argued that as the share of agriculture in
national income and employment diminishes with
economic development, the inequality of incomes
tends to increase because the industrial sector as
opposed to the agricultural sector is characterised
by economies of scale and hence promotes larger
sized production units (Ahluwalia, 1976,p.321).
Such economies, it is often said, do not exist in
agriculture and for this reason agriculture-induced
growth as in Punjab would be regarded as promoting
equality in the long run. The logical basis for
this argument is weak, however, and it may well be,
as Marxist scholars have often argued, that with the

development of capitalism in agriculture, economies
of scale also appear and in time lead to sharp
differentiation of the peasantry and hence increase
in inequality in the long run. Some of the issues
involved in this debate are discussed, in the
context of prospects for capitalist agricultural
development in Punjab, in Chapter VII.

We shall turn now to the second group of
issues, namely those relating to the relative
position of agricultural labour. This problem has
often been discussed in terms of the standard
two-factor neo-classical model with competitive
factor markets and disembodied technical change. In
such a model, whether or not inequality increases
with economic growth depends on (i) the relative
change in factor supplies, (ii) factor bias of
technical change, and (iii) the possibilities of
substitution between factors. Assuming land and
labour to be the only factors, the supply of labour
to rise faster than that of land (with population
pressure reducing the land/man ratio) and technical
change to be land augmenting (as in the new
technology of irrigation-seed-fertilisers), labour's
share will rise/fall, hence relative inequality
fall/rise, accordingly as the elasticity of
substitution is greater/less than unity
respectively. Substitution possibilities in
agricultural production are generally believed to be
high and the elasticity of substitution above unity.
If so, inequality should decline with technical
progress in agriculture. While the model lends
itself easily to formulation in terms of empirical
data, its assumptions appear to be unrealistic. The
assumption of competitive markets, for example, may
well fail to hold for our region especially in
respect of labour (Chaudhri, 1979). The restriction
to two factors neglects capital-land substitution in
agriculture; subsuming capital under land begs the
question, while introducing all three factors (or
more if intermediate inputs are included in the
production function) makes the conditions for
labour's share to rise or fall extremely difficult
to interpret. Finally, the estimation of the
magnitude of the elasticity of substitution (which
is crucial to the argument) raises a series of
conceptual difficulties (Morawetz, 1974), and these
turn on the question of the measurability of capital

in the context of the neo-classical production function, which was discussed earlier (Chapter II). For these reasons, we do not believe such an approach to be particularly useful. The model of economic development with surplus labour, its well-known limitations notwithstanding, may still give more useful insights on this particular question. This assumes that the elasticity of supply of labour with respect to the real wage rate is very high if not infinite. This could be due to of demographic pressure together with the lack of opportunities for industrial employment, migration of labour from nearby regions with significantly lower agricultural wage rates, and the differentiation of the peasantry, with marginal farmers, erstwhile tenants and village artisans joining the ranks of landless labourers. In such a situation the rise in the real wage of agricultural labourers is likely to lag behind increases in labour productivity. What happens to labour's share of total income would then depend on the nature of the change in the demand for labour.[6] The impact of mechanisation becomes of crucial importance in this context. It was concluded in Chapter II that while the direct effect of mechanisation was to replace labour, its joint effects along with new varieties, higher fertiliser use and irrigation would probably be to increase labour use per unit area. However, given that the real wage may be expected to rise only slowly, labour's share would depend on the demand for labour per unit of output which would almost certainly decline over time.

So far, the argument has been essentially in terms of the supply of output. The elasticity of demand for agricultural products could also be relevant. If this demand is price-inelastic, a lower price of agricultural goods due to increase in labour productivity achieved by the new technology would tend to reduce labour's share. In the post-Green Revolution period, government intervention in the market has generally acted to prevent increases in agricultural productivity from bringing about any significant decline in the relative price of agricultural products. These are good political reasons for this (Mitra, 1978). But, if the relative price of agricultural goods should actually rise and the demand for agricultural

products is indeed price-inelastic, the same
argument implies that labour's share should
increase. The assumption of an inelastic demand for
agricultural output may itself turn out to be
unjustified, however, if exports are considered as
well as domestic demand. The argument becomes much
more complex if shifts in demand as well as supply
curves are allowed for.[7]
    The incidence of inequality depends on both the
rate of economic growth and the way in which growth
is achieved; but it may also be affected by
political organisation and the nature of the state,
social customs such as caste, the legal system and
in particular laws governing the inheritance of
agricultural land, and social and economic policies
pursued by the government. Here, we shall comment
briefly only on the last of the factors
mentioned.[8] That redistributive policies can never
succeed, at least in the initial stages of economic
growth, is implied by some exponents of the
inverted-U hypothesis who conclude from their
regression that income distribution must become
worse before it can get better. No regression
analysis, irrespective of the 'goodness of fit', can
prove any such proposition.[9] Moreover, the studies
concerned show considerable variation between
countries in respect of the extent to which actual
inequality exceeds or falls short of its predicted
value; such variation may well be due to
differences in policy, a point to which we shall
return in our concluding chapter.

2 Trends in Average Levels of Consumption

During 1960-61 to 1979-80, per capita real income in
Punjab increased more than threefolds which suggests
a significant rise in levels of living.[10] Private
real consumption per head appears to have risen much
less; but in assessing how much it did the
following considerations must be kept in mind. The
standard of living of poor people consists largely
in what they eat and this could be judged by such
criteria as the content of their diet and the extent
to which it provides protection against disease and
malnutrition and satisfies consumer tastes and
socio-cultural norms. As long as levels

of nutrition are low, it is the average energy
(calorie) intake that is of primary concern; but
beyond a certain minimum level, it is the other
aspects which are more important.[11] In rural
Punjab the average calorie intake was already fairly
high by the early 1960s. This level appears to have
been maintained during the decade of the 1960s.
Later, especially after the mid-1970s, it increased
further. The average current calorie intake in
rural Punjab is believed to be around 3000 - a level
commonly regarded as a characteristic of more
developed countries, though unlike in such countries
in Punjab it can still vary sharply over good and
bad agricultural years.

More important, however, have been changes in
the food basket. Here, the most striking increase
has been in the consumption of milk and milk
products both in terms of their share of the total
expenditure on food and in absolute terms. The
average consumption of liquid milk in rural Punjab
rose from about 7 litres per capita per month in
1961-62 to 13 litres in 1974-75. The consumption of
sugar, vegetables, fruits and nuts and eggs also
increased significantly. While the proportion of
total food expenditure on and total calories derived
from cereals has diminished, within the cereals
group itself coarse grains have been partly replaced
by wheat. On the debit side, the per capita
consumption of pulses has diminished due to higher
relative prices. On the whole, a rather more
balanced and palatable diet is now enjoyed by rural
Punjabis, at least by the farmers.

Shelter is usually regarded as coming next only
to food and perhaps clothing in the hierarchy of
wants. In the Punjab context where, because of
their traditional connections with the army, many
peasant families have long been relatively
well-clothed, an improvement in housing conditions
is probably a more sensitive indicator of a rise in
the standard of living than better clothing. Both
the quantity of housing (the number of dwelling
units) relative to the number of rural households
and its quality, as reflected in such
characteristics as the average number of persons per
room, the kind of material used in construction, the
type of structure and the amenities available, are
relevant in this context. According to the Census

data, the number of dwellings in rural Punjab as a
percentage of the number of households increased
only marginally from 65.4 in 1961 to 66.3 in 1971;
and while the number of persons per room increased
from 2.46 to 2.68 indicating an increase in
overcrowding, in other respects the quality of
housing improved significantly. The proportion of
rural houses with walls constructed predominantly of
durable materials went up from about one-third in
1961 to over half in 1971, and during this period
the proportion of kutcha houses declined from 79 to
47 per cent. Census data on housing for 1981 have
not yet been released, but some evidence from other
sources suggests that conditions improved much
faster during the 1970s.[12] Punjab and Haryana
together with Manipur and Sikkim were the only
states reporting no shortage in rural housing in
1980 [13]; and facilities such as electricity,
toilets and kitchens have also improved, even though
they are still primitive.

So far our discussion of living standards has
been concerned exclusively with rural areas, where
over three-quarters of the population of the region
live. We now turn to the quarter of the population
who are urban. Their calorie intake per head, which
had long been low - not only much lower than the
rural average but also below the nutritional norms -
seems to have improved a little during the 1960s but
is still inadequate. Improvements in housing were
more significant. During 1967-71, while the gap
between the number of urban households and that of
residential houses changed relatively little, and
the average number of persons per room marginally
increased, the quality of housing improved greatly:
the proportion of kutcha houses, for example,
declined from 63 per cent to a mere 8 per cent, and
there was some improvement as well in the provision
of amenities such as sanitation, water supply, and
electricity. Higher levels of living are often
reflected in associated changes in demographic
indices, for example, an increase in life-expectancy
or a decline in infant mortality; and in turn such
indices, the expectation of life at birth in
particular, have been widely regarded as indirect
indicators of the standard of living. In our
analysis of demographic trends in Punjab in an
earlier chapter, we have already pointed out the

pervasive links between economic and demographic variables. Improvements in nutrition, housing conditions and environmental hygiene and spread of female education were found to be particularly important in bringing about the decline in mortality, especially that of infants, and the increase in the expectation of life at birth, which have occurred in Punjab especially since the mid 1960s.

3 Trends in Inequality in Punjab

The basic estimates on inequality in Punjab are presented in Tables 6.1 and 6.2. There are two sets of estimates: on household per capita consumption (Table 6.1) and on household land (Table 6.2). The figures of Table 6.1 suggest a changing pattern in the inequality of consumption which is shown graphically in Figure 6.1. First, there are sharp year to year fluctuations in inequality in the pre-Green Revolution period, but these are much less pronounced in the period since 1965. Secondly, inequality clearly declined after 1960-61 and also after 1965-66. These results contradict the Kuznets hypothesis as well as the received wisdom that the Green Revolution led to sharp increases in rural inequality in Punjab.[14] On the contrary, a negative association between inequality and growth is suggested by the fact that the simple correlation coefficient between the Gini coefficient (G) and the index of agricultural production [15], taking all the years for which values of G are available, is -0.52, which is statistically significant at the 5 per cent level.

Inequalities in rural income and consumption are expected to be closely related to those in the distribution of land. This distribution itself may, however, be specified in a number of alternative ways. It may, for example, refer to all rural households, i.e. including the landless, or to rural households with land, i.e. excluding the landless, the former distribution being necessarily more unequal. Again, because of the possibility of leasing in and leasing out land, the distribution of owned land may be different from that of the area operated. If it is the smaller farmers who tend to

Table 6.1 INEQUALITY IN CONSUMPTION IN PUNJAB
1957-58 TO 1973-74

Year	Household Consumption (Rural) Gini-Coefficients
1957 - 58	.323
1959 - 60	.303
1960 - 61	.369
1961 - 62	.354
1963 - 64	.302
1964 - 65	.323
1965 - 66	.330
1966 - 67	.308
1967 - 68	.298
1968 - 69	.278
1970 - 71	.297
1971 - 72	-
1973 - 74	.290

Source: Ahluwalia (1964)

Table 6.2 INEQUALITY IN OWNERSHIP AND OPERATION OF
LAND IN PUNJAB 1953-54 TO 1971-72

Year	Gini-Coefficients			
	All Rural Households		Rural Household excluding Landless	
	Owned Area	Operated Area	Owned Area	Operated Area
1953 - 54	.764	.715	.634	.606
1960 - 61	.766	.747	.731	.710
1971 - 72	.767	.745	.744	.442

Source: Computed from data given in NSS (16th, 17th
and 18th Rounds).

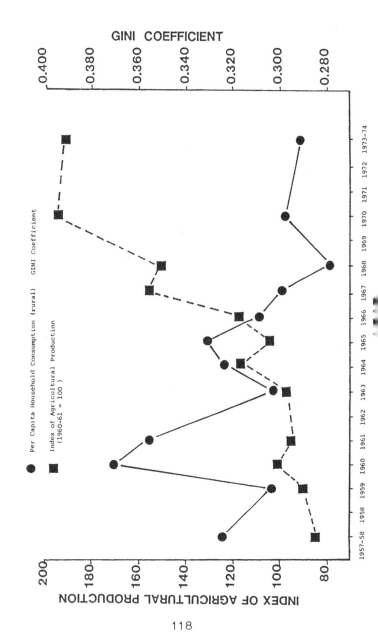

FIGURE 6.1 TRENDS IN AGRICULTURAL PRODUCTION AND INEQUALITY IN PUNJAB , 1957-58 TO 1973-74

● Per Capita Household Consumption (rural) GINI Coefficient

■ Index of Agricultural Production
  (1960-61 = 100)

lease in land from those with larger ownership holdings, the distribution of owned land will show a higher level of inequality than that of the area operated. If, on the other hand, the smaller farmers, on balance, lease out because, for example, they find cultivating small patches of land no longer rewarding, it is the distribution of area operated that will show greater inequality.

Estimates of the Gini coefficient for all rural households and for households other than landless households, are given in Table 6.2 separately for owned and operated area. These show a very high but nearly constant level of inequality in land ownership for all rural households. As far as the households with land are concerned, there is an increase in inequality in owned area, especially in the 1950s.[16] This reflects the absence of radical land reform, and perhaps also the success of measures undertaken during the early 1950s for the consolidation of holdings, which led to transfers of land from owners of small parcels to others. The inequality in area operated by such households is lower than that in ownership throughout the period and, after an initial increase, it shows a sharp decline during the 1960s. This suggests both that leasing arrangements have, on balance, exerted an equalising influence between farmers of different size groups, and that this effect has been much more pronounced in the later years. This may, indeed, be largely responsible for the decrease in consumption inequality after the mid-1960s.

Our results are subject to errors of measurement; for example, measures of inequality given in Table 6.1 relate to the distribution of households by household expenditure per capita, so that while variations in needs arising from differences in the size of households are adjusted for, those due to differences in age and sex composition are not. This could be done by computing inequality indices with respect to the distribution of households by household expenditure per adult equivalent, an appropriate equivalence scale being constructed for this purpose, and it has been done by Satya Paul (1983) for Punjab state, 1970-71. But time-series of such indices are not yet available.

A more important source of bias is that the NSS

data on consumption expenditure from which our
indices of consumption inequality are derived are in
current prices: comparisons of inequality in real
consumption based on such figures could be
misleading if relative prices have changed. Poorer
households tend to spend a larger share of their
budget on necessary items such as food, the demand
for which is price-inelastic. Hence, if their
prices rise relatively to luxuries over the period,
measures of inequality derived from current price
expenditure data would overstate the decline (or
understate the rise) in inequality of real
consumption.[17] We do not know how important this
was, and there are no econometric studies to help
us. Some writers have pointed to the faster price
rise as compared to wheat or coarse grains such as
jowar and bajra which figure prominently in the poor
man's diet, a trend which can be attributed to
technological change associated with the Green
Revolution itself.[18] The effects of this have been
mitigated, however, by the substitution of wheat for
coarse grains in the consumption basket[19],
although this could also reflect a decline in the
consumption of an inferior commodity due to an
increase in real income. An important question is
how far the price of food items as a whole increase
relative to other items of consumption. Here the
evidence is of a small upward trend, together with
sharp year-to-year fluctuations.[20] The decline in
the inequality of consumption in real terms was
therefore somewhat less than the estimates of Table
6.2 suggest.
　　Certain problems also arise from the way in
which NSS data were collected. For example, a
multiplicity of prices have been used in evaluating
consumption, especially that of cereals. NSS
schedules distinguish between four types of
household consumption: (a) home-grown stock, (b)
cash purchase, (c) receipts in exchange for goods
and services, and (d) free collection. These have
been valued respectively at the ex-farm price, the
actual purchase price, the average retail price
prevailing in the locality during the reference
period, and the average producer price during the
reference period. The ex-farm price, which is free
of middlemen's profits and other trading margins,
tends to be lower than either the actual purchase

price or the average retail price. This results in
the cereals consumption of cultivating households,
which is largely out of their home-grown stock,
being undervalued as compared to the consumption of
non-cultivating households who either purchase it
for cash in the market or receive it in exchange for
labour services. Inequality in consumption as
between cultivating and non-cultivating households
is thus understated.

Again, the NSS sample probably suffers from
under-representation of the rural rich, who are few
in number, both because of their inaccessibility to
NSS investigators and because of the failure of the
sample design to use such techniques as inverse
sampling or the stratification of the population by
income levels.[21] This is another possible source
of a downward bias in the estimate of inequality.

The food consumption of those households in the
top income groups who are included may have been
overestimated by including food given to servants
and to agricultural labourers employed by them;
this may account for implausibly high figures for
calorie intake (5800 or more per capita per day).
On the other hand, there seems also to be an
underestimation of the consumption of the poor
because of (i) incomplete coverage of home-grown
produce and (ii) the omission of gifts of left-over
food and used clothes received from employers or
others from the 'free collections' category of
consumption.[22] Both these errors tend to overstate
the extent of inequality.

To sum up, the net effect of these errors is
probably that the NSS data somewhat understates the
true extent of inequality; and this could
relevant in comparing the level of inequality in
Punjab with that elsewhere. However, we are not
convinced that the errors described changed in a
systematic way over time.[23] Hence, our judgement
that the trend in inequality especially since the
mid-1960s was downward stands.

4 Trends in Poverty and the Composition of the
Rural Poor

So far we have been concerned with the relationship
between economic growth and inequality in Punjab,

and as we have seen the relationship is fairly complex. In this section we shall look briefly at the relationship between economic growth and poverty. For this purpose we take poverty in its conventional sense of the proportion of the relevant population whose consumption is below a specified minimum level of subsistence. Problems involved in this definition are discussed in the Appendix D. Intuitively, one would expect the relationship between poverty and economic growth to be fairly straightforward. Continued economic growth should lead to a rise in the level of real per capita consumption, and this, in turn, to diminished poverty. However, neither of these two conditions need necessarily hold.[24] The processes bringing about a higher rate of economic growth could themselves be such that either an increase in real output per head was not accompanied by any sustained rise in real consumption per head, or even if it was its distribution tended to become cumulatively more unfavourable to the poor. In either case, the expected negative relationship between economic growth and poverty could fail to be realised. Did such conditions prevail in Punjab?

In post-independence Punjab the growth rate of agricultural output showed considerable fluctuations from year to year, but over the period as a whole agricultural output as well as total output, both in absolute and in per capita terms, recorded an impressive rise especially in the post-Green Revolution period. For real consumption per head, the evidence is more indirect and qualitative in nature, but it still points unmistakably to a significant rise, especially for the rural population, though its magnitude was clearly lower than that of the rise in output per head. As regards the extent of inequality, we have already noted that the Gini coefficient for rural household consumption per head also showed sharp year-to-year fluctuations, especially during the earlier part of the period, but clearly declined from the mid-1960s. On the basis of these considerations, it would be reasonable to expect that the incidence of poverty in Punjab should also be subject to short-term variations but should decline over time, especially since the mid-1960s. That such an expectation corresponds roughly to what actually happened can be

seen from the available empirical evidence on
poverty in Punjab, summarised in Table A.4.

Our first comment on these figures concerns
year-to-year fluctuations in poverty. According to
Table 6.1 the proportion of the Punjab rural
population in poverty declined in a single year by
more than five percentage points on one occasion
(1959-60 to 1960-61) and by nearly ten percentage
points on another (1967-68 to 1968-69). It
increased in one two-year period (1961-62 to
1963-64) by about seven percentage points and in
another by more than seven. Such drastic
fluctuations in the extent of rural poverty could be
due to variation between good and bad agricultural
years,[25] more generally to structural reasons,
common to many less developed countries which allow
wide fluctuations in the price of staple foods [26],
and perhaps to purely statistical reasons.[27]

As far as the overall trend is concerned,
during the decade 1967-68 to 1977-78 rural poverty
clearly declined. The common view that the Green
Revolution has made no dent on rural poverty in
Punjab is thus quite mistaken.[28] But neither does
the Punjab experience provide a basis for relying on
a high rate of agricultural growth alone as an
automatic cure for rural poverty. As we have seen
earlier, Punjab experienced rapid agricultural
growth during the 1950s, but the poverty ratio
remained high. A sustained decline in rural poverty
did not occur till after the late 1960s and a hard
core of poverty still persists.

So far we have discussed poverty in terms of a
low level of consumption among households, and there
were good reasons for doing so. Our earlier
discussion emphasised that both small and large
farmers have gained from improved production
possibilities in Punjab, and it is this that has led
to a reduction in poverty among farmers, which we
believe to be both genuine and irreversible.
Agricultural labourers, who for all practical
purposes have no land or cattle, belong to a
separate category, however; and it is they who form
a majority of the rural poor in Punjab.[29] Their
income, derived predominantly from wage-labour,
depends crucially on the level of output and the
technology of production. How far it has increased
in real terms since the mid-1960s is discussed

below. Trends in real wages and in employment will be discussed in turn. As regards the first, our discussion is based on index numbers of money and real wages of agricultural labour in Punjab in harvesting, which can be taken to represent the yearly peak wage.

Money wages have increased greatly over the period. The most rapid increase occurred in the initial years, between 1965-66 and 1969-70. During the first three years of the 1970s wages increased only slowly, but later they rose sharply and the upward trend continues.

Trends in real wages are more difficult to interpret. Not only did prices, especially of food grains, rise over the period, they fluctuated sharply from year to year, and index number problems are involved in adjusting money wages appropriately.[30] Index numbers of real wage were derived by using four different price indices. Two are variants of the CPIAL, the first being an index of food prices alone, while the second also includes other items of consumption. The index of Punjab Wheat Farm Harvest Price yields a wheat equivalent measure of the real wage. The Murthy and Murthy index as mentioned earlier is a fractile specific index based on the consumption mix of the lowest three deciles per household per capita expenditure.

For the first few years of the period, trends in real wages computed from different price indices are fairly similar, but afterwards they diverge. The CPIAL indices, for example, show a sharp decline in real wages in harvesting from 1972-73 to 1973-74, the Murthy and Murthy index shows a sharp increase, and the wheat price index a massive increase. Estimates of the rise in real harvesting wages during 1965-66 to 1973-74 are approximately 3 per cent, 8 per cent, 84 per cent and 89 per cent, depending on whether we use the CPIAL Food, CPIAL General, the Wheat Farm Harvest, or Murthy and Murthy price index respectively.

One cannot point to any one of these index numbers as being clearly superior to the others, and none satisfied the strict conditions required of a 'true' cost of living index. However, as a measure of the rise in real agricultural wages in Punjab the CPIAL, for reasons stated earlier, tends to suffer

Table 6.3 EXTENT OF DEPENDENCY AMONG AGRICULTURAL LABOUR HOUSEHOLDS IN PUNJAB

	1950-51	1964-65	1974-75
1. Average size of households	4.90	5.50	5.71
2. Average number of wage earners per household	1.70	1.77	1.86
3. Average number of male earners per household	1.30	1.37	1.40
4. Average earning strength per household	1.70	1.85	2.25
5. Dependency Ratio:			
(A) = 1 - (2)/(1)	.65	.68	.67
(B) = 1 - (3)/(1)	.73	.75	.75
(C) = 1 - (4)/(1)	.65	.66	.61
6. Percentage of agricultural labour households to all rural households	7.7	14.3	15.8

Source: National Sample Surveys, various rounds.

from a conservative bias. On the other hand, an equivalent measure in terms of wheat, which had over a long period a lower rate of price increase than most other cereals, would be over-optimistic. The Murthy and Murthy index has the merit of being based on a weighting system directly related to the consumption of the rural poor, although unlike the others it is constructed from All India rather than Punjab data, but it is not available after 1973-74.

In our judgement, real wages of agricultural workers in Punjab have probably increased much more than indicated by the CPIAL, the index which has generally been used in the present context; but even so, they have lagged far behind the rise in real agricultural output.

The real income of agricultural labourers depends also on their level of employment, which we have already discussed in Chapter II. We found that the data, incomplete though they were, suggested a decline in employment as measured by the average number of full days of wage-paid work per year done by an agricultural labourer. However, although the average size of agricultural labour households has increased over time, their dependency ratio has not (Table 6.3), which suggests that the trend of employment for the household may have been slightly more optimistic than for the individual labourer.

5 Agricultural Growth, Poverty and Inequality in Punjab

We have found that post-independence agricultural growth in Punjab led to an improvement in levels of living, and that at least since the mid-1960s inequality and poverty have decreased. It was argued that the conclusion stands despite limitations of our statistical estimates. Two further qualifications are pointed out below.

One is our neglect of the consumption of non-market goods and services provided by the public authorities. Our estimates of inequality were derived from data on private household consumption expenditure and so were estimates of poverty. During the last two decades, there has been a considerable increase in the supply of public goods and services in the Punjab, such as primary health services at the village level, elementary education, drinking water, sanitation, and nutrition programmes including free school meals for children. These were often provided with the professed aim of benefiting the rural poor. We have no quantitative studies of how far these aims were in fact achieved. However, our earlier discussion in Chapter III suggests that at least in some areas, especially in health and education, substantial benefits accrued to the poor, although those of rural road and electrification programmes went mainly to richer farmers. On the whole, the effect of publicly supplied goods and services is likely to have helped to improve the conditions of the poor and reduce inequality in real consumption. They have also promoted mobility and skill formation, and more generally helped in redressing inequalities of opportunity, and they have contributed substantially towards improving the quality of life in Punjab, especially for its rural population.

A more fundamental limitation of our empirical analysis in this chapter is its overriding concern with consumption. Poverty, for example, is defined in effect as a low consumption level and inequality discussed mostly in terms of the inequality of levels of consumption. For a deeper understanding of the relationship between growth, inequality and poverty and its implication for the future of Punjab, we must go beyond the analysis of consumption patterns and look as well at the role of different social groups in the production process. This has already been emphasised in our critique of the Kuznets hypothesis in the first section of this chapter; it underlies our subsequent discussion of the position of agricultural labour, and we shall discuss it more specifically in the next chapter in terms of the political economy of the development process.

NOTES

1. Kuznets (1966, pp.424-25) points out that this arises largely through the income shares of the very top ordinal groups being higher in less developed countries; the shares of the much larger low ordinal groups are not significantly lower.

2. The myth of an egalitarian village
community dies hard despite much evidence to the
contrary provided by historians, cf. Hilton (1973,
p.32): 'The peasant community was not a community
of equals. The stratification of peasant
communities, moreover, was at least as old as the
earliest records which we have of them. This
suggests that such polarization of fortunes as there
was between village rich and village poor could not
simply have resulted from competition in production
for the market, important though this factor was
from time to time in generating social
differentiation'.

3. See Chapter II.

4. Such a trend has also been reported by
studies of other Asian and Latin American countries.
A good survey of their results is given by Cline in
Frank and Webb (1977).

5. Cf. Cline op.cit, p.320: 'The Green
Revolution appears to have concentrated rural income
not because of any scale requirements of the new
technology but because of inequality in the
distribution of the new inputs involved. To the
extent that this inequality is associated with
demand based on farmer decision making (rather than
supply factors associated with credit and other
conditions of access to inputs), the historical
pattern of eventual adoption of technology by small
producers after initial adoption by larger ones
suggests that the concentration may be transitory'.

6. Cf. Cline, op.cit, p.32: 'An important
distributional aspect of the Green Revolution
mentioned by several authors is the expulsion of
former tenants by landowners desiring to operate the
land themselves in view of the new profitability of
production with improved inputs. With land value
enhanced by increased profitability, the owner can
no longer afford the chance that tenants will
acquire legal claim to it, and at the same time the
owner is prepared to incur risks associated with
production expenses that formerly were unattractive
compared to the simpler option of permitting
operation by sharecroppers'.

7. For an analysis, in these terms, of the
agricultural development in the United States, see
Schultz (1953).

8. For a discussion of the political and legal

aspects of inequality in India see Beteille (1982).
9. Kuznets himself states clearly that 'Low income inequality change in the process of a country's economic growth could only be answered for growth under defined economic and social conditions', and recognises that his own analysis deals with the specific experience of the currently developed countries 'which grew up under the aegis of business enterprise' (1955).
10. See especially Table 8.4 .
11. Similar trends were experienced elsewhere in the process of development. See for example Scholliers and Vandenbroeke (1982).
12. A number of micro studies confirm the improvement in housing conditions suggested by state level statistics. Leaf's (1983) report on one such study, based on intensive surveys of a particular village in Ambala district once in 1965 and again in 1978, finds 'a substantial improvement in housing and shelter, resulting in more work areas, more storage facilities, better sanitation and a general increase of comfort and level of welfare' (p.237).
13. Estimates of housing shortage in each state, separately for rural and urban areas, have been prepared by the National Buildings Organisation (NBO), Government of India. This is conceived of as the shortfall of the existing from the desired number of houses, the latter being estimated on the basis of specified norms, viz. that every household should have a housing unit to itself which may be either pucca or semi-pucca in urban areas, but could be either of these or the serviceable kutcha category in rural areas (these terms are defined in the Glossery). The NBO reports the shortage of rural houses both in Punjab and in Haryana and of urban houses in Haryana to be nil: while the shortage of urban houses in Punjab is put at 60,000. (See Handbook of Housing Statistics, National Buildings Organisation, Government of India, 1980, Table 3.10, page 24). The concept of a housing shortage underlying the definition used might not be so relevant in a society where the number of separate households itself depends on the number of dwellings available, but rural society in Punjab, characterised by joint rather than nuclear families, is not of this kind.
14. For example, Griffin (1981).

15. Using the official index for all crops (Chapter II, Table 2.1).
16. Statistical problems involved in the measurement of land inequality are discussed in a number of papers in Indian Journal of Agricultural Economics, Conference Number, October-December 1980.
17. Conceptual problems involved are discussed in Muellebauer (1974) and in Deaton and Muellabauer (1980, pp.176-78).
18. See for example, Saith (1981).
19. Tyagi (1982) compares the distribution of cereal consumption for agricultural labour households in India 1956-57 with that of consumer expenditure classes up to Rs.43 per month at 1973-74 prices for the year 1973-74, a roughly comparable group. His calculations show that while the proportion of expenditure on coarse cereals (jawar, bajra, ragi and small millets), barley and gram has declined, that on wheat has increased from about 7.5 per cent to about 16.8 per cent. The corresponding change in Punjab is likely to have been even more pronounced.
20. An interesting attempt to correct the Gini coefficient of consumption in Punjab for differential price change as between different fractile groups was made by Rajaraman (1975). The correction was done by deflating consumption expenditure of the different groups by separate price indices with 1960-61 a base. The households were ranked in order of per capita consumption (derived by dividing household consumption expenditure by the family size) and a price index was computed separately for each of three different fractile groups of the resulting cumulative distribution, viz. the poorest (30%), the middle (30-74%), and rich (74% and higher). The Lorenz Curve was derived accordingly. However, even though she finds that for every commodity group other than cereals, price rose less for the higher groups, the final value of price indices for 1970-71 were very similar, namely, 209.96, 206.83 and 206.26 respectively for the poor, middle and the rich groups, which is explained by the greater weightage of cereals in the consumption of the poor. In consequence, her actual measure of inequality in 1970-71 is affected only marginally by the price

index used. But this may not be so for years when food prices rose sharply, for example 1966-67 and 1974-75.

21. Dandekar and Rath (1971), Rudra (1974), Sukhatme (1970).

22. According to Boserup (1981) the underestimation of the 'free collection' component of food consumption is a general phenomenon. 'In many areas hunting and gathering contribute significantly to the calorie and protein intake of agricultural populations. Usually this food is underestimated or forgotten in calculations of food supply and nutritional levels' (p.41).

23. See, in this connection, Bardhan (1974).

24. Indeed, the contrary has sometimes been asserted, for example, by Adelman and Morris (1973), who conclude on the basis of a step-wise variance analysis applied to an inter-country cross section that 'development is accompanied by an absolute as well as a relative decline in the average income of the very poor'. That their result is not statistically robust has been shown by Cline (1977); but a more fundamental criticism is that they fail to identify causal processes which could lead to such an outcome. Cf. our remarks in section 5 of this chapter.

25. The relevant variation is between good and bad years in the all- India, not just the Punjab, context: food scarcity in Eastern India in 1967-68 contributed towards raising food prices, and hence poverty, in Punjab. This is due both to market forces and to the transfer of food grains from surplus to deficit areas through the public distribution system.

26. Cf. Deaton and Muellbauer (1980, p.178).

27. See Appendix D.

28. Some writers by comparing the magnitudes of the poverty ratio as between two particular years have even argued that poverty in Punjab has been increasing, e.g. Bardhan (1974) for 1960-61 and 1967-68; Rajaraman (1975) for 1960-61 and 1970-71, but since the poverty ratio shows considerable year to year fluctuations, conclusions based on such two-point comparisons can be misleading. Others have emphasised that the coefficient of regression of the poverty ratio on time estimated by Ahluwatia is not statistically significant, indicating in

their view that poverty has not declined (e.g. Griffin, 1981). However, what this really shows is the failure of poverty to decline smoothly with respect to time, rather than the absence of a decline as such. Indeed, if we replace the figures of Table 6.1, by three-year moving averages and consider the regression of these averages on time over the years 1966-67 to 1970-71, we find that this regression coefficient is negative and highly significant.

29. According to Minhas (1970) only a third of the rural poor in India were agricultural labourers and more than half small or marginal farmers, but Bardhan (1970) and Dandekar and Rath (1971) regard the former as the more important component. There is considerable regional variation in this respect, however, and most scholars agree on the greater importance of agricultural labour among the rural poor of the Punjab/Haryana region. Further, the line demarcating marginal farmers from agricultural labourers is far from distinct since many of the former supplement their income by hiring out their labour from time to time.

30. See our discussion in Appendix D. An offsetting factor should be noted however. Agricultural labourers, especially at harvest time, have traditionally received part of their wage in kind, and its cash equivalent is included in the statistics of money wages underlying Table A.6. There are indications, however, that the 'kind' component has been valued at less than its market price and also that its relative importance has been reduced, usually at the insistance of employers. Computed trends could therefore overestimate the real rise in wages. Cf. Lewis (1976): 'An index of mass welfare which took in not just the output of wage-goods but also such things as the number of hospital beds per thousand or the percentage of population living within half a mile of a public water tap, would show that the poor have gained much more from development than it is now fashionable to believe'. This is certainly true of Punjab.

Chapter VII

GROWTH AND STRUCTURE: EMERGING PATTERNS

No social order ever perishes before all the productive
forces for which there is room in it have developed and new
higher relations of production never appear before the material
conditions of their existence have matured in the womb of
the old society itself.

Karl Marx

That agricultural development in Punjab has been
truly remarkable not only by comparison with the
period before independence but by any standards is
evident from our work so far. This does not,
however, guarantee that such growth is viable in the
sense that it is likely to continue in future.
There are numerous instances of economies which
had apparently 'taken off' on the basis of a spurt
of industrial growth but which subsequently reverted
to stagnation [1], and this may occur in the case of
agricultural growth as well. How far, then, does
the growth that has occurred provide a basis for a
self-sustained economic development in Punjab? An
attempt to answer the question leads directly to
issues connected with the structure of agriculture
and of the Punjab economy as a whole. In its
simplest terms, the link can be described as
follows. The process of rapid growth in a
traditional society, of the kind that has occurred
in Punjab, tends to affect not just the level of
output but both the structure of productive assets
and the socio-economic relationships in which
different groups of the village population stand
vis-a-vis one another and tò those outside the
village. These, in turn, could have an important
bearing not only on the degree of social and
political stability but also on the pattern of
inter-temporal resource allocation in general and
rates of investment and innovation in particular,
and hence on the prospects of future growth.
A proper assessment of the long-term future of
the Punjab economy cannot therefore be based simply

133

on statistical time-series analyses, however refined, of the past.[2] One must try to understand what kind of economic structure has emerged in Punjab and what its 'laws of motion' are likely to be.

The chapter is organised as follows. The first section examines how far tendencies towards capitalist development have appeared in Punjab agriculture and looks in particular at the process of differentiation of the peasantry. In second section we discuss what kind of sectoral structure of output and of employment has emerged in Punjab, and the structure of the manufacturing sector in particular is briefly discussed in the last section.

1 Agricultural Growth and Differentiation of the Peasantry

There has been a lively debate in India on whether a capitalist 'breakthrough' has occurred in Punjab agriculture. This is clearly of some importance if one tries to answer the question posed in the introduction to this chapter: how far economic growth in Punjab has led to structural change and how far this growth is viable?

To this matter we now turn, and we start by taking up some of the more well-known features that are associated with the development of capitalism in agriculture. Among these the most important areas are: (i) an expansion of productive forces in agriculture; (ii) continuous technical change leading to high rates of increase in productivity; (iii) the reinvestment of surpluses generated in agriculture for capital accumulation; (iv) production of crops for exchange rather than for subsistence of farmers themselves; and (v) the differentiation of the peasantry.

The penetration of capital into Punjab agriculture was described in some detail in Chapter II. Within a relatively short time period, there has been a massive increase in the stock of real capital in agriculture, especially in farm equipment and machinery which have become increasingly modernised. Capital per unit of land has increased greatly and to a lesser extent so has capital per unit of the labour force (which itself showed

134

considerable increase). Agricultural output increased even faster than capital.

Pre-capitalist modes of production tend to be conservative in the sense that they provide no built-in stimulus tending to bring about a systematic improvement in productivity. Under capitalism, given a minimum degree of competition, the quest for profits provides such a stimulus. In consequence, there are more inventions, and they are adopted faster. Some constraints imposed by conditions specific to agriculture- for example the importance of soil and climate, the nature of the crop cycle and the interdependence of different agricultural operations- cannot be eliminated whatever the mode of production might be. The pace of technical progress in agriculture even under the most favourable circumstances tends to be slower than that of industry.

The achievement of continuous technical change under capitalism is closely linked to the use of surpluses generated in production for reinvestment. The dynamics of capitalist agriculture,in particular,have been attributed to the reinvestment of profits by capitalist farmers making possible a continued process of capital accumulation. In Punjab, as we have seen, the mechanical component of technical progress was no less important than the 'biological' one; while there is not much direct evidence on how such investment was financed, the reinvestment, especially by bigger farmers, of surpluses generated in agricultural production appears to have played a major role.

Production for the market, as distinct from subsistence farming, had existed in Punjab agriculture for centuries; but only in the case of cash crops and only for big farmers had it been really important. With the increased profitability of cultivation, and greater access to grain markets due to improvements in transport, storage, credit and information, production for the market has also come to play a dominant role in the case of food crops, especially wheat and rice, and for a great many small farmers of Punjab as well as for the bigger farmers.

Finally, we consider the question of the differentiation of the peasantry which turns, essentially, on the nature of the relationships between different groups or classes involved in the production process, referred to at the beginning of this chapter. The concept of agrarian transition provides a useful framework for such an analysis. Punjab has long been known as a land of peasant agriculture; as noted earlier, the peasantry was far from homogeneous. With the penetration of capital in agriculture, the stratification of agricultural producers usually begins to develop in a more definite way. In a developed capitalist agriculture the essential division is between a class of capitalist farmers who are in a position to exercise control over the means of production and a class who have only their labour-power to sell, i.e. a landless 'rural proletariat'. Certain aspects of the development of such a group in Punjab have already been described in the previous chapter.

The development of capitalist agriculture itself, according to the theory of transition, is a very long process. What the theory requires is the beginning of a process of rural class differentiation, which could set up a tendency towards the kind of overall development that it postulates. Whether or not such a process is at work in Punjab agriculture can only be tested by a systematic empirical analysis of peasant stratification. Our earlier discussion in Chapter V of the farm-size productivity relationship is clearly relevant here, but the focus of that discussion was different. As we noted there, the available data do not permit us to be systematic, and we shall only offer a few remarks based on studies already cited and, more specially, on the comprehensive recent survey of agricultural conditions in Punjab state in 1974-75 by Bhalla and Chaddha (1981). Data given in the Bhalla and Chaddha study enable us to distinguish between five different size groups of farmers, viz. those operating less than 2.5 acres (I); those operating more than 2.5 but less than 5 acres (II); those operating 5 to 12.5 acres (III); those operating 12.5 acres or more (IV); and those operating 25 acres or more (V), a sub-group of (IV). The cut-off points for demarcation have no particular sanction but are chosen so as to correspond roughly to the groups that have been often described in the Indian context as marginal, small, middle, big and very big farmers.

Comments on this data are offered below. We start by looking at differentiation between the farm-size groups in respect of the availability of land. The average size of holdings (net operated area) for big farmers was about twelve times as much as for marginal farmers and two-and-a-half times that for middle farmers. Since family size tends to rise with farm size, differences in the household tend to be less than those in the average size of holdings. But even on a per capita basis the big farmers had access to more than double the amount of land than middle farmers had and about eight times as much as the marginal farmers. There was considerable differentiation in respect of farm machinery. As many as 69 per cent of the very big farmers owned tractors as against 36 per cent of the big farmers, 7 per cent of the middle farmers, and 3 per cent of small farmers, while marginal farmers owned no tractors. Nearly nine-tenths of the biggest farmers, three-quarters of the big farmers, nearly three-fifths of the middle farmers, a third of the small farmers and just one-sixth of the marginal farmers owned tubewells. The inequality in the ownership of equipment was only a little offset by the hiring of machinery services, since just a third of the marginal farmers and less than a quarter of small farmers used such services.

Differences in the use of material inputs were much less pronounced. Expenditure per acre on seeds was more or less the same in different size groups. In the case of manures and fertilisers, while expenditure per acre rose with farm size as in other studies, the differences were fairly small. The only important items for which expenditure per acre increase sharply with farm size were diesel and electricity, reflecting the inequality in the ownership of tubewells. On the other hand, expenditure per acre on draught cattle declined sharply with farm size (because many large farmers have switched over to tractors while smaller farmers have too little land to permit efficient utilisation of bullock capacity), and the same was true for hire charges, which is to be expected since only the smaller farmers who lack equipment need to hire.

We do not attach any particular significance to the precise numerical results cited above. But the

orders of magnitude are not too different from what
other studies suggest, and they do allow two
important qualitative conclusions to be drawn.
First, Punjab farmers are not a homogeneous group.
Here (as elsewhere) the notion of a single 'rural'
class has little reality. Secondly, the differences
between different groups are not so much in respect
of input use, crop mix, or yield rates, but in the
basic land and capital assets held. This come out
most sharply if we look at the marginal farmers.
That they attempt to maximise their returns from
agricultural activity subject to the constraints
imposed by limited land is sufficiently clear. It
is shown, for example, by their relatively high
level of use of current inputs such as chemical
fertilisers and seeds of high yielding varieties, by
the fact that their crop mix is almost identical
with that of the bigger farmers, by their higher
cropping intensity as compared to the other groups,
and by their intensive application of human labour.
The availability of canal irrigation has enabled
them to adopt the new technology of production. As
a result, they have succeeded in achieving fairly
high yield rates per acre, often the same as the
overall average. Nevertheless, these rates are not
high enough to compensate for the basic limitation
of little land. Income per head is low, farm
business income per capita a fraction of the overall
average. Hence, their desperate search for
alternative occupations. Indeed the major part of
the total household income appears to be derived
from non-farm sources of which the most important
are remittances and pension, dairying and poultry
production, agricultural and non-agricultural wage
employment, and household enterprise. Cultivation
itself increasingly becomes a peripheral activity as
far as they are concerned. This may lead them
eventually to opt out of cultivation altogether, as
indeed some of them appear to have already done. A
comparison of the NSS data on the size distribution
of the operational holdings for 1953-54 and 1971-72
(Chapter VI) shows, for example, that of all the
farms operating less than 2.5 acres the number of
those operating less than one acre decreased not
only relatively but also absolutely; in consequence
the average size of holding in the group as a whole
(i.e. 0-2.5 acres) went up.

We shall conclude this discussion by mentioning certain aspects of peasant differentiation that were not considered above.

An important, some would say the most important, mark of differentiation is in respect of the different uses of labour-power, viz. self-employment, hiring in labour and hiring out one's own labour to others. How one relates to the buying or selling of labour-power could be used to define one's 'class' position. Capitalist farmers, for example, depend predominantly on hired labour and do not work for others. At the other end of the spectrum, landless labourers are neither self-employed nor employers of hired labour but are themselves hired. Other groups would come in between. It would be useful to know how far a classification in terms of the pattern of labour use and one according to the size of holdings correspond. From scattered data and from personal observation, we have the strong impression that in this region boundaries between different categories of farmers in respect of labour use are not at all clearly drawn. Both 'small' and 'middle' farmers, for example, hire in labour and hire out their own labour in addition to being self-employed and even marginal farmers may hire in labour on some occasions. On the other hand, even the largest farmers work on their farm, not only in a supervisory capacity, reflecting the weakness of caste-based taboos against manual work which have helped to maintain social distance between them and farming families elsewhere in India. Some possible implications of such 'fluidity' for future developments are touched on in the concluding section of this chapter.

Secondly, there is the question of money-lending, which some have interpreted as a semi-feudal relic and others as a major constraint on the growth of agrarian capitalism. That money-lending has been of exceptional importance in this region is true enough. The Punjab Banking Enquiry Commission of 1929 described money-lending to agriculturalists as Punjab's largest industry. Concern about the social and political consequences of rural indebtedness in Punjab led to the enactment of a series of legislative measures starting with the Punjab Alienation of Land Act (1900) and

culminating in the Act of 1938. While these had the
effect of curbing the activities of professional
(non-agricultural) money-lenders, they encouraged
the growth of a new class of agriculturalist
money-lenders. Since independence the role and
functions of such money-lending have changed
substantially under the impact first of the massive
growth of facilities for institutional credit, and
secondly of technological advance which made farming
itself far more profitable. Nevertheless
money-lending remains a lucrative enterprise.
However, the view of money-lending as a constraint
on the capitalist development of agriculture is
strangely lopsided; its impact will depend not only
on the terms and conditions under which loans are
incurred but more especially on the purposes for
which they are spent. Since the 1960s a predominant
part of the loans incurred by larger farmers and a
considerable part even of those by small farmers
have been for productive purposes, for example for
buying chemical fertilisers, pesticides or farm
equipment. More recently, cases of larger farmers
borrowing institutional credit on concessional terms
and lending part of it to smaller farmers at higher
rates of interest have been reported by some
observers as becoming increasingly common. It would
be odd to describe this as retarding the spread of
capitalism in agriculture.

Thirdly, we have neglected differences arising
from tenancy. That it has led to land use being
less unequal than land ownership was noted in the
previous chapter. Reasons for believing that in
Punjab the tenant-owner difference is not of crucial
importance are given below.

Even in principle the distinction between
ownership and tenancy of land is one of degree.
Each represents a customary bundle of rights and
obligations related to land use, and customs have
changed over time under the impact of technology and
law. The rights and obligations involved are also
quite different as between the two principal kinds
of tenant, viz. an occupancy tenant who enjoys
security of tenure as long as he pays the rent and
can usually pass it on to his male descendants on
the same terms, and a tenant-at-will who enjoys no
such rights, whose term of tenancy is one year or
less and who can be easily evicted. Occupancy

tenants, who accounted for over half the cultivated area under tenancy in Punjab in 1953-54, are not easily distinguishable from peasant proprietors in economic terms, though they might rank lower in the social hierarchy. The position of tenants-at-will is quite different and it is about them that the economic theory of tenancy has been essentially concerned.

There is a large theoretical literature on the allocative efficiency of tenurial arrangements, especially of share-cropping, as compared with owner-farming, but results are highly sensitive to the assumptions used. However, the relevance of the theory is limited by its static nature and by its failure to take into account that decisions about hiring labour, land and even equipment may be inter-linked.

More relevant for us are behavioural theories of tenancy which attempt to make contingent predictions about the future; for example, Rao (1976) points out a possibility under share-cropping of a conflict of interest between a (risk-neutral) landowner and a (risk-averse) tenant. The landowner would favour a choice of crops and techniques leading to maximise expected net revenue; the tenant, one with a lower expected revenue and a lower risk. Share-cropping would therefore be more likely where substitution possibilities between inputs or crops were lacking, so that there would be little scope for decision-making. Contrary-wise, with the emergence of a new agricultural technology, which expands the scope for decision-making by providing a wider range of choice, the incidence of share-cropping should decrease. This is, in fact, what has happened. According to NSS data, while in 1953-54 just over 40 per cent of the cultivated land in Punjab was under tenancy, by 1970-71 the proportion had gone down to only 25 per cent or so. This was due largely to a decline in share-cropping tenancy, which had traditionally been the single most important type of tenurial arrangement in this region. However, the decline may have been only partly due to the reasons just described. Perhaps a more important reason was land reform. Throughout the 1950s in Punjab, as elsewhere in India, masses of tenants were evicted by landowners in order to use land for self-cultivation and so escape the imposition of land-ceiling acts. Since that time, despite new and 'hidden' forms of tenancy coming up in some parts of the region, on the whole tenancy has diminished substantially in importance and we believe will continue to do so in future. Tenant farmers of the region also do not appear, by and large, to be less efficient than owner farmers of the same size group either in allocating resources or in adapting innovations. Relationships between different sections of the agricultural community, and more generally the viability of agricultural growth, depend not only on what is happening within the agricultural sector but on trends in the structure of production as a whole. These are discussed in the next section.

2 Sectoral Composition of Output and Workforce

The process of economic development is reflected directly in changes in the structure of production. As the work of Kuznets (1965) has taught us, such changes occur rather slowly and are therefore best studied in long historical contexts, consisting of five to six decades. Unfortunately data available for the Punjab economy for this purpose does not cover even two decades. We do not, therefore, expect to observe a great deal of structural change. However, the magnitudes and direction of changes in sectoral shares of real output and employment even for a short period of 15 years (1960-75) should provide some clues about the long term prospects of economic development in Punjab. Value added per male worker, in both the agricultural and non-agricultural sectors of the Punjab economy, increased substantially over the period 1961-71, as is clear from Table 7.1.

The rate of growth of labour productivity in the non-agricultural sector was thus well over twice that in the agricultural sector, despite the Green Revolution. Changes in relative (male) labour productivities of the two sectors during the period are summed up in Table 7.2.

The gap between agricultural and non-agricultural labour productivity widened during 1960-61 and 1970-71. This is reflected in the change in the labour productivity of the non-agricultural sector relative to that of agriculture from 1.091 to 1.464.

Had the sectoral shares of employment remained the same in the two years, differential productivity growth would have resulted in a fall in the share of the agricultural sector in the State Domestic Product (SDP). How far in fact there was a shift in the structure of the labour force from the agricultural to the non-agricultural sector, will be examined in the next section.

Sectoral shares of value added at constant prices are reported in Table 7.3 and Figure 7.1. Agriculture in 1960-61 contributed 59.2 per cent of the State Domestic Product. Its share declined to 53.8 per cent in 1965-66 but rose to 56.5 per cent

Table 7.1 VALUE ADDED PER MALE WORKER IN PUNJAB
(1960-61 Prices)

	1960-61 Rs	1970-71 Rs	Percentage Increase
Agricultural Sector	1052	1443	37
Industrial Sector	1148	2112	84
All	1089	1675	54

Source:Reserve Bank of India Bulletin April, 1978.

Table 7.2 LABOUR PRODUCTIVITY RATIOS

	1960-61	1970-71	1971/1961
Agricultural Sector to Statewide	0.966	0.862	0.892
Non-Agricultural Sector to Statewide	1.054	1.261	1.196
Non-Agricultural Sector to Agricultural Sector	1.091	1.464	1.346

Source: Computed from RBI Bulletin April, 1978.

1970-71. By 1974-75 it had gone back on the
expected declining trend and was 51.1 per cent.
Over the period 1960-61 to 1974-75, there was thus a
decline of 8 percentage points in the share of
agriculture in total product measured at constant
prices, and this was distributed almost equally as
increases in the manufacturing and the service
sectors. Major changes also occured in the relative
shares of some industries within the manufacturing
and service sectors. Within the service sector,
public administration in value added increased by
350 per cent over a 15-year period. Both in
absolute value and as a percentage of total product
it is fairly small; yet its supportive role for
agricultural development through provision of
infrastructure, public and merit goods discussed in
Chapter III is unmistakable. Services such as
education, health, agricultural research, extension
and general development administration are included
in this category. Similarly, transport and
communications, banking and insurance, and
electricity, gas and water supply registered
substantial increases in their levels of output as
is clear from Table 7.4.

Since the commercial banking system in India
was nationalised in 1969 and financial institutions
like cooperative credit societies are also included
in the category of banking, a substantial part of
the banking and insurance is essentially a public
system activity. Similarly, electricity, gas and
water supplies are also public system activities
undertaken by either the state government or local
bodies. In transport and communications it is only
the goods transport component using private trucks
which is a private sector activity, the rest of it
is managed by the state. In this sense all the
fast-growing modern sectors which supported growth
of the agricultural sector are largely run and
managed by the state and form part of the public
sector. Their growth essentially represents a
supportive service to the rest of the economy and,
in particular, to production in the agricultural
sector.

Other sub-sectors which registered substantial
expansion in their value-added contribution to the
State Domestic Product in real terms are
manufacturing, construction, trade, hotels and

Table 7.3   STATE DOMESTIC PRODUCT AT FACTOR COST
BY SECTOR OF ORIGIN (1960-61 Prices)
(Per cent)

Sector	1960-61	1965-66	1970-71	1974-75
Agriculture	59.200	53.828	56.474	51.055
Manufacturing	20.321	23.859	22.246	25.297
Services	20.158	22.316	21.222	23.649
Total	100.000	100.000	100.000	100.000
State Domestic Product (Rs Million)	6278.7	7352.3	10685.0	11819.0
Index	100	117	170	188

Source: Reserve Bank of India Bullettin, April 1978.

145

restaurants, and real estate and ownership of dwellings as is clear from Table 7.5

As Table 7.5 brings out, there was substantial increase in output in manufacturing and trade, both of which grew much faster than total State Domestic Product. Construction, on the other hand, hardly kept pace with the increases in SDP, the index number of the former increasing only from 100 to 180 while in the SDP it was from 100 to 188. Real estate and dwellings did not register any major increase; their index increased from 100 in 1960-61 to a mere 123 in 1974-75. Thus the widely held belief that much of the fruits of the Green Revolution have gone into construction and real estate is not supported by this evidence, although it is possible that some of the activity in this sector may have gone unrecorded.

Shifts in the labour force are much smaller than in output and they contain a number of surprises. Since the detailed sectoral classification followed in the Census of population differs from that used in national and state income computations, per worker productivity cannot be computed for each of the sub-sectors separately. When dealing with the 'lower level' sub-sectors, we shall therefore restrict ourselves to commenting on trends in output and in employment separately.

After 1951, with increases in the output and productivity in the agricultural sector, symptoms of structural change slowly started appearing. The sectoral allocation of male workers in the labour force for the years 1951-81 is reported in Table 7.6. A major expansion is shown in the number of male workers in the labour force and, since agriculture was the dominant sector and the development process in the Punjab economy also started in the1950s in this sector, it is natural that employment opportunities should arise in the first instance in agriculture. The percentage share of male workers in agriculture has been between 61 and 65 per cent, which is roughly similar to the historical pattern since 1881. In this respect there is no empirical evidence of structural change. We observed in Chapter II that there were different sub-periods of accelerated and decelerated growth of agricultural output. The acceleration of agricultural growth has enabled agriculture to

Table 7.4 CHANGES IN VALUE-ADDED SHARES OF IMPORTANT SECTORS SUPPORTING AGRICULTURE (1960-61 Prices, Rs Millions)

	1960-61		1965-66		1970-71		1974-75	
	Amount	Index	Amount	Index	Amount	Index	Amount	Index
Electricity, Gas and Water Supply	46.0	100	81.4	177	134.0	291	152.3	331
Transport and Communications	244.1	100	316.4	130	461.7	189	587.4	241
Banking and Insurance	68.5	100	90.8	133	142.3	208	154.1	225
Public Administration	151.1	100	248.2	164	348.5	231	529.1	350
State Domestic Product	6278.7	100	7352.3	117	10685.0	170	11819.0	188

Source: Computed from RBI Bulletin, April 1978.

attract or retain a large proportion of the labour
force, e.g. because of the spread of irrigation
facilities in the early 1950s and the Green
Revolution in the mid 1960s.

The situation with respect to other sub-sectors
is rather different. The trade and commerce and
transport and communications sub-sectors have
attracted larger numbers of workers in comparison to
the growth of total labour force. The index of
numbers employed in trade and commerce increased
from 100 in 1961 to 175 in 1981, the largest
increase occurring between 1971 and 1981. The
labour force in transport and communications also
increased by 50 per cent over two decades. Both
these sectors seem to have benefited by the growth
of the agricultural sector.

As against these, the male workforce in
construction shows a declining trend, having
decreased by 18 per cent over 20 years. We pointed
out earlier that in traditional activities such as
agriculture, reported female participation rates
were remarkably unstable and that for this reason
analysis using magnitudes of the male workforce only
would give more sensible results. Even in the
non-agricultural sector, reported female
participation rates appear to be unstable. We have
analysed details of the female workforce for these
sectors separately. In construction the index of
the number of female workers declined from 100 to 33
over 20 years. Female employment in the
manufacturing sub-sector appears to be even more
unstable. Due to changes in the definition of
workers between Censuses which affected the female
workforce more than the male workforce even in the
non-agricultural sector, it is not worthwhile to
analyse these figures without a great many
adjustments which we could not undertake. The only
general comment that we can make is that the Punjab
non-agricultural sector is certainly not following
the patterns observed in the British history of
economic development where employment of females
(especially in the textile industry) increased
substantially during the Industrial Revolution.

Employment in the manufacturing sector at best
has grown more slowly than the growth of the labour
force as a whole. Within the manufacturing sector
there seems to be a major shift from household to

Table 7.5 VALUE-ADDED SHARES OF SELECTED SUB-SECTORS
(1960-61 Prices, RS Millions)

Sub-sector	1960-61		1965-66		1970-71		1974-75	
	Amount	Index	Amount	Index	Amount	Index	Amount	Index
Manufacturing	683.9	100	995.3	146	1351.5	198	1706.9	250
Construction	296.0	100	357.1	121	424.4	143	533.5	180
Trade, Hotels and Restaurants	580.6	100	695.8	120	1057.0	182	1260.1	217
Real Estate and Dwellings	139.1	100	149.4	107	160.7	116	170.9	123
State Domestic Product	6278.7	100	73523.3	117	10685.0	170	11819.0	188

Source: Computed from RBI Bulletin, April 1978.

factory type employment. The index of household
employment declined from 100 to 63 during 1960-61 to
1970-71 while that of non-household sub-sectors
increased from 199 to 219 giving a net overall
increase of employment at a rate lower than the
labour force increase.

In Table 7.7 we present index numbers of value
added shares and employment of male workforce in
important sub-sectors of the Punjab economy between
1961 and 1981.[3] This table is derived from data
used in Tables 7.1 to 7.6 with some interpolation
for 1965-66 and 1974-75. Sectoral classifications
for value added and labour force in the
manufacturing sub-sector are also slightly
different[4], and in interpreting this table these
adjustments and differences should be kept in mind.
Total value added increased faster than the male
labour force giving us increasing per worker
productivity in the economy of the state as a whole.
This decidedly is a sharp break from the situation
observed in Chapter I for the period before 1951.
In this sense a growth process in the Punjab economy
has certainly started. For sustained economic
development to occur this growth has to be
'substantial' and continue over a long period.
Moreover, increases in per capita as well as per
worker output are required. In the Punjab economy
such increases have been substantial; whether these
can be sustained is not certain yet. In the
agricultural sector value added increased by 62 per
cent over 15 years while the labour force increased
by only 28 per cent for the same period leading to a
substantial rise in per worker productivity in the
agricultural sector.

Value added in the manufacturing sector
increased by two-and-a-half times over this period,
while labour force employed in this sector increased
by only 9 per cent. This is a substantial increase
in per worker productivity which contributed to the
widening gap between the agricultural and
non-agricultural sectors noted earlier. It may be
attributed to capital deepening, to the use of more
skilled and educated workers in this sector and to a
changing output mix in the sector itself. Nothing
much can be said about the emerging trend except
that employment in this sector in recent years seems
to be picking up as well, as is clear from the

Table 7.6 SECTORAL ALLOCATION OF MALE WORKERS IN PUNJAB 1951 to 1981

Occupational Categories	1951		1961		1971		1981	
	Number	Index	Number	Index	Number	Index	Number	Index
Total Male Workers	3,137,459	55	5,764,363	100	6,380,094	111	7,896,160	137
Agricultural Sector	2,120,922	63	3,352,413	100	4,177,978	118	4,866,101	138
Manufacturing (Total)	247,859	33	759,468	100	683,070	90	973,409	128
(a) Household	n.a.	n.a.	420,558	100	200,930	48	262,905	63
(b) Organised	n.a.	n.a.	338,910	100	482,140	142	710,514	210
Construction								
Males	31,374	23	135,172	100	122,966	91	110,808	82
Females	n.a.	n.a.	5,071	100	3,292	65	1,672	33
Trade and Commerce	269,821	71	380,663	100	496,245	130	664,305	175
Transport and Communication	69,024	49	141,611	100	173,048	122	215,746	152
Manufacturing								
HH	n.a.	n.a.	120,291	100	189,799	158	24,044	20
Fac	n.a.	n.a.	15,148	100	13,538	89	7,177	47
Total	n.a.	n.a.	135,439	100	203,337	150	31,221	23

Note: 1951 Census occupational classification is not strictly comparable to that of the later
Censuses. Figures reported here for 1951 are adjusted as per percentages in table 1.3.
Source: Census of India, various reports.

increase in the index of labour force in this
sub-sector from 109 in 1974-75 to 128 in 1981.
Moreover, this increase, as noted above, is in the
non-household manufacturing sub-sector. Since this
sub-sector largely employed wage-labour, cost
consciousness and profitability considerations
should lead to further capital deepening.

The index of value-added by construction
increased from 100 to 180, roughly similar to that
in the State Domestic Product. The labour force
employed in this sector actually declined; the
index moved from 100 in 1961 to 87 in 1974-75 so
that productivity rose substantially. The decline
in employment is probably due to the fact that the
technology of construction has undergone significant
changes with the introduction and use of labour-
saving machinery such as concrete mixers, floor
polishers, etc. At the same time, use of
factory-made construction material is replacing
construction and preparation of this material on
construction sites.

Value added by trade and commerce also
increased much faster than the State Domestic
Product as a whole. The size of the labour force
also followed a similar pattern. This is indicative
of rising labour productivity at a rate similar to
that of the economy as a whole. Value added in the
transport and communications sub-sector increased
much faster than did employment. In index number
terms value added rose from 100 to 241 while
employment increased from 100 to 137. Per worker
productivity in this sub-sector, as in
manufacturing, has thus increased substantially.
This is partly due to a shift away from traditional
methods of transporting of agricultural produce in
bullock carts to use of tractor trollies and partly
due to use of larger-capacity, long-haul trucks
which has made loading and unloading at too many
points no longer necessary. This is indicative of
capital deepening in transport which, besides
raising labour productivity, is likely to strengthen
a trend towards integration of markets for
agricultural output and purchased modern inputs.

Table 7.7   INDEX OF VALUE-ADDED SHARES AND LABOUR-FORCE SHARES

	1960-61	1965-66	1970-71	1974-75	1980-81
Total Value Added	100	117	170	188	137
Total Labour Force	100	105	110	124	
Agricultural					
Value Added	100	106	162	162	138
Labour Force	100	109	118	128	
Manufacturing					
Value Added	100	146	198	250	128
Labour Force	100	95	90	109	
Construction					
Value Added	100	121	143	180	82
Labour Force	100	96	91	87	
Trade and Commerce					
Value Added	100	120	182	217	175
Labour Force	100	115	130	152	
Transport and Communication					
Value Added	100	130	189	241	152
Labour Force	100	110	122	137	

Source: Reserve Bank of India Bulletin, April 1978.

3 Emerging Patterns in the Structure of the Punjab
Economy

The manufacturing sector of Punjab in 1971 employed
about 17 per cent of the labour force and
contributed 22 per cent to the State Domestic
Product. It is a small but expanding sector with a
high rate of growth in value-added per worker. The
manufacturing sector is by no means homogeneous and
therefore overall expansion and growth of
productivity in this sub-sector is very unevenly
distributed.
     The economic organisation of this sub-sector
consists of three types of units, namely household
industry, unregistered workshops, and registered
factories. Their respective numbers and the
proportion of employment in each type are reported
in Tables 7.8 and 7.9. Registered factories,
although they are small in number, account for half

Table 7.8 MANUFACTURING ESTABLISHMENTS IN PUNJAB,
          1971

	Punjab	Haryana	%	Total
Household Industry	36683	24109	22	27792
Unregistered Workshops	62275	248446	74	92397
Registered Factories	3660	1676	8	5336

Source: Gupta(1982).

Table 7.9 EMPLOYMENT IN MANUFACTURING SECTOR 1971
           (Per Cent)

	Punjab	Haryana
Household Industry	16	18
Unregistered Workshops	42	29
Registered Factories	42	53
Total	100	100

Source: Gupta (1982).

the employment and about 60 per cent of the share of value-added. Unregistered workshops are second in order of importance and household industry, which is largely concentrated in rural areas, has a less than 20 per cent contribution to value added and employment.

Industries constituting the bulk of manufacturing sector activity are food products, edible oils, textile products, cotton and wool textiles. These are agro-based industries and they account for half of the manufacturing sectors' production and slightly more than half of industrial employment. Linkages of these industries with the agricultural sector are reasonably strong, both regionally and nationally. Industries producing non-metallic mineral products, machinery, machine tools and parts, transport equipment and parts and other manufacturing and repair industries provide for the other half of value added and slightly less than half of manufacturing sector employment. Value-added per worker in these industries is higher than that of agro-based industries, so is the fixed capital per worker and capital output ratio.[5] Backward linkages of these industries within the region are weak while relatively stronger with India as a whole.

Forward linkages with the region are strong. This pattern of industries, as correctly observed by Gupta (1982), is indistinguishable from the all-India pattern. Three points are worth noting in the industrial sector of Punjab. First, Punjab does not have any significant mineral deposits. Therefore mineral-based industries do not seem to have a major potential. Secondly, Punjab does not have any large scale organised industry and the prospects of establishing one in the face of competition from states such as Gujarat and Maharashtra do not seem good. Third, a large number of small establishments consisting of unregistered workshops and household units have access to electric power. Compared with the availability of skilled manpower these units have considerable promise if the institutional climate is right.

NOTES
1. Argentina is cited as one such example.

See Maddison(1970).

2. Political economy of economic growth is a complex process. History has shown that the interaction of social and economic change rarely proceeds smoothly. In fact the speed of economic change significantly affects social and political institutions.

3. We have discussed the problem of stability of mearused participation rates for female workers in chapter II. Due to the problems of reliability and comparability of data on occupational classification of female workers we have restricted our analysis to male workforce only.

4. Sectoral classification of the manufacturing sector for value added is different from that used in the Census for occupational classification. We have combined catagories to make them broadly comparable.

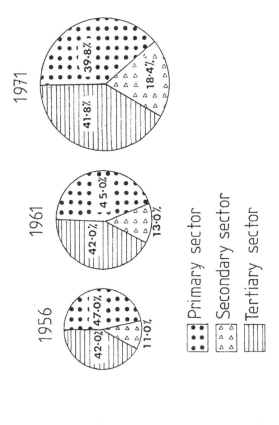

FIGURE 7.1 VALUE OF NET OUTPUT IN PUNJAB STATE, 1956 to 1971

SOURCE Statiatical abstracts of Punjab and Haryana, various issues.

Chapter VIII

AGRICULTURE AND THE DEVELOPMENT PROCESS

*Development, therefore is not just a harmless and peaceful
process of growth like that of organic life but a hard
and obstinate struggle with itself.*

Hegel

This study is concerned with agricultural growth and
its impact on the development process in Punjab. We
have examined the magnitude of growth rates of
agricultural output and the increases in the supply
of productive factors which together with technical
progress made high growth rates of output possible
(Chapter II), contributions made to growth by state
policy and social infrastructure (Chapter III), the
inter relationship between economic growth and
demographic transition (Chapter IV), fragmentation
and integration of the market (Chapter V), the
impact of growth on rural inequality, poverty and
standard of living (Chapter VI), and its impact on
economic structure (Chapter VII). We shall conclude
our study by looking at the contribution of
agricultural growth in Punjab to the development
process as a whole. The chapter consist of four
sections. In the first, we analyse the nature of
the contribution of Punjab agriculture to the
development process. This contribution is then
compared to that of agriculture in Japan in an
earlier period, and to the experience of other
Indian states more recently. The final section
looks at some possible futures.

1 Contribution of Agriculture to the Development
Process

Following Kuznets, we may distinguish between three
aspects of economic development: aggregative,
structural and international. The first refers to

158

increases in total and per capita real product; the second to shifts in the relative importance of various productive sectors on the one hand and of forms of economic organisation and economic classes on the other; the third to the interrelations of a nation's growth with that of others through international trade, finance and communications. The three aspects are linked. A rise in per capita product for example, by changing patterns of consumption and savings, would affect the structure of production; and it is the development of new industries and new methods of production which normally bring about a structural shift and makes it possible for a rise in per capita product to occur. Agricultural growth can contribute to any or all of these aspects and according to Kuznets there are again three ways in which it can do so. First it can contribute to the growth of a nation by growing rapidly itself. This may be called its product contribution. The extent of this contribution depends on the share of agriculture in total output at the beginning of the process and on the relative rates of growth in agricultural and non-agricultural output. However, as long as the rate of growth in non-agricultural sectors is higher than that in agriculture, the proportional contribution of agriculture to the growth of total product is bound to decline over time. How fast this will happen depends on how much more productive the non-agricultural sector is as compared to agriculture and on the extent of migration from one to the other.

Secondly, there is the market contribution of agriculture which it makes by purchasing productive inputs and consumer goods from the industry and services in exchange for its own product. In the process agriculture helps other sectors to emerge and develop and so helps to diversify economic structure. At the same time, agriculture itself becomes integrated with the rest of the economy.

The third contribution of agriculture consists of transfers of resources to other productive sectors through taxation or by using loans or savings originating in agriculture to finance the growth of other sectors or by a shift of labour from agriculture to other sectors.

The Kuznets taxonomy provides a convenient way

of looking at the contribution of agriculture to economic development; but in applying it to the Punjab experience, at least two important limitations must be kept in mind. First Kuznets was concerned with the economic growth of nations. Punjab is a region. The contribution of Punjab agriculture, whether it is a product, market or factor contribution, could be defined in regional or all-India terms. In keeping with the focus of this study, it is the regional interpretation that we shall discuss below unless the context indicates otherwise.

Secondly, the prices that a Punjab farmer receives for his crops are subject to regulation by government, so are those of inputs such as fertiliser, diesel, and electricity; and the prices of farm machinery are not only determined in a highly oligopolistic market, but are also affected by the nature of government policies in respect of protection, fiscal policy and foreign exchange control. Hence the distinction between a market contribution and a factor contribution is not a hard and fast one. There is also a more general point, viz. that the extent of the contribution of agriculture to economic development is not determined by agricultural performance alone.

The direct or product contribution of Punjab agriculture to the economic growth of Punjab has been substantial, both because agricultural output had a high growth rate and because it had a high initial share of total Punjab output.

The market contribution of Punjab agriculture has also been quite substantial not only in a regional but also in an all-India framework. The use of chemical fertilisers, machinery and equipment and mechanical power in place of manures, draught animals and hand-made tools has led to a higher share of purchased inputs in gross agricultural product. A greater degree of integration of agriculture with the rest of the economy is the result. The point is not simply that agricultural activity is accompanied by other kinds of production. In the traditional agriculture of Punjab, spinning and weaving were carried out as secondary domestic occupations of the agricultural households themselves, but they were not technically necessary for agricultural production, nor did they involve much market contribution. On the other hand, as Marx emphasised, the penetration of capital in agriculture has the consequence that the use of improved seeds, pesticides, chemical fertilisers as well as implements and machinery become an integral part of the agricultural activity itself.

The other important component of the market contribution of Punjab agriculture is its marketed surplus, and several quantitative studies have established that its magnitude is quite considerable.

The factor contribution of Punjab agriculture has been little. There is no tax on agricultural income. In 1979-80, land revenue for Punjab and Haryana together accounted for less than 10 per cent of total revenue. Indirect taxes on agricultural goods are also light. We have little or no information on the extent to which savings of agriculturalists have financed capital formation elsewhere in the economy, but all indications are that this must have been negligible. Finally, labour did not move out of agriculture, rather the contrary.

2 Growth of Agriculture in Punjab (1950-1980) and Japan (1880-1935)

It may be of interest to look at the agricultural growth of Indian Punjab and its role in the development process in a comparative framework. For this purpose we shall present two separate comparisons. In the present section, recent Punjab experience will be compared with that of Japan in the past. In the next, we shall compare it with what has been happening in some other regions of India during the same period.

If we wish to look at the agricultural growth of the Punjab in an international perspective one fruitful comparison is with Japan, which is often presented as a model of success in a land-scarce economy. However, comparing the economic development of post-independence Punjab with that of a contemporary post-World War II Japan would be obviously inappropriate. A more relevant comparison is with Japanese economic growth during 1868-1935. The fact that Japan is a country while Punjab is a

Agriculture and the Development Process

region of India clearly affects the interpretative
significance of such a comparison. As we have seen,
the impetus of growth in Punjab after partition had
an all-India dimension, and this dimension is
relevant in drawing policy conclusions about
regional growth. We need to remember as well that
Punjab had the advantage of a late start, giving it
access to the Green Revolution technology, farm
machinery and chemical fertilisers in the early
stages of its agricultural development which were
not available to Japan in 1868.
    While these differences are important, there is
still sufficient similarity between the two
situations to make a comparison interesting. The
total population as well as the percentage share of
the total workforce that was engaged in agricultural
and allied activity in the Japan of 1880, the
earliest year for which reliable estimates are
available, are very similar to those of Punjab in
1951. Both faced a land-scarce situation. Both had
small-scale peasant agriculture, with fairly similar
relative levels of share tenancy. In both cases the
state played an important initiating and supportive
role, and in both agricultural development was
achieved through an interaction between state
policies and the response by individual producers to
these and to price incentives. In neither case was
there major land reform during the period, in Japan,
this came after World War II. In both cases the
rate of agricultural growth was high. In both the
break from past trend rates of agricultural growth
was unmistakable. In Japan, output increased much
faster after 1880 than either before the Meiji
Restoration of 1868, or during the transition of
1868-80.
    In our discussion of Punjab's agricultural
growth in Chapter II, it was observed that while the
trend rate of growth of agricultural output was
high, there were sub-periods of accelerating and
decelerating growth. The same was true in Japan
though the duration of such periods was longer in
Japan. Ohkawa and Rosevsky (1968) divide Japan in
this period into three sub-periods:

              (i) 1868-85, a period of transition
              dominated by institutional reforms and
              rising labour input per acre resulting in

slowly accelerating growth;

(ii) 1885-1905, a period which itself consisted of two parts: GDP rose at an annual rate of 4.3 per cent during 1885-1898 and only at 2.3 per cent during 1898-1905;

(iii) 1906-1930, a period of another upswing which peaked in 1919, the annual rate of growth during 1906-1920 being 4.2 per cent. It was lower but still positive during 1920-1930.

According to Ohkawa (1965) the basic historical features which Japanese agriculture manifested before arriving at a turning point around 1930 are as follows:
  (i) the labour force in agriculture remained the same although its relative share of the total labour force declined;

  (ii) the acreage of arable land also remained nearly the same providing for a stable land man ratio;

  (iii) owner farmers remained the core of the land system and the distribution by farm size changed little;

  (iv) returns to scale remained neutral; and

  (v) inputs of working and fixed capital increased considerably, accompanying an increase in output and labour productivity.
Punjab agriculture's features and trends reflect very similar patterns as is clear from Table 8.1.
  One important difference between agricultural trends in Punjab in 1951-1972 and in Japan in 1880-1905 is that the land-labour ratio slightly increased in Japan while in Punjab it declined. This could be attributed to the fact that the labour force in Punjab agriculture increased much faster -at an annual compound rate of 1.4 per cent as compared to only 0.1 per cent in Japan. The shift out of agriculture had already started in

Table 8.1   SOME COMPARISONS BETWEEN PUNJAB (1951-1972) AND JAPANESE
AGRICULTURE (1880-1935)

	Punjab		Japan		Japan	
	Base Year 1951	Terminal Year 1972	Base Year 1880	Terminal Year 1905	Base Year 1880	Terminal Year 1935
Population (millions)	12.6	23.9	36.6	46.6	36.6	68.7
Per Cent in Agriculture (%)	66	64.1	75	67.1	75	47
Land Productivity Y/Ln	100	247	100	125	100	184
Labour Productivity Y/La	100	206	100	149	100	267
Capital Productivity Y/K	100	168	100	125	100	132
Land/Labour ratio Ln/La	100	83	100	102	100	103
Growth of Area (Compound)	0.6		0.9		0.5	
Growth of Labour Force Agriculture (Compound)	1.4		0.1		-0.2	
Growth of private Fixed Capital (Compound)	7.6		11.0		11.0	
Growth of Value Added in Agriculture (Y) (Compound)	5.1		1.8		1.6	

Source: For Japanese historical data, see Hayami et al. (1975) and
Ohkawa and Shinohara, ed. (1979). Punjab data and ratios are
our own calculations.

Japan while in Punjab it has not occurred so far.
There were also some other historical factors at
work in Japan, for example, migration for colonial
settlement. The effect of these could be seen from
the changes in output per worker and in output per
unit of area sown in the two regions. In Punjab
output per hectare increased much faster than
output per worker. In Japan the reverse was true.

Another important difference is in respect of
the growth rate of private fixed capital in
agriculture which was much higher in Punjab than in
Japan. Yet output per unit of capital increased
faster in Punjab. Essentially, this is because the
growth of value added in agriculture was much
higher in Punjab.

If the comparison is between Punjab 1951-1972
and Japan 1880-1935 rather than Japan 1880-1905,
two important differences emerge. In the earlier
period (1880-1905), the agricultural labour force
in Japan had increased slightly, at a rate of 0.1
per cent per year. After 1905, it declined.
Structural changes in Japan which had started in
the last two decades of the nineteenth century seem
to have accelerated in the first three decades of
this century. The land-labour ratio remained
stable; the index fell by only one point in 30
years. The contrast with Punjab in this respect is
clear in terms of trends towards structural change.

Another major difference is the advantage of a
late start which Punjab had. Punjab's higher rates
of growth of agricultural output, as compared to
Japan, can be attributed to the High Yielding
Varieties which made the Green Revolution possible
and to the availability of a whole range of modern
agricultural machinery and implements: the small
size of holdings and the predominance of rice
cultivation may have acted as deterrents to
mechanisation in Japan. It is perhaps because of
this that part of the historical experience of
Japan over a period of 60 years has been telescoped
in Punjab in only two decades.

The most important differences between
Japanese and Punjab agriculture are the fiscal
arrangements and financing of state-provided
infrastructure. In Japan before the 1868 Meiji
Restoration, according to estimates of Tsuru (1953)
about 37 per cent of agricultural output went to
the feudal lords and their retainers. After the
major land reforms of 1873 about 34 per cent of
output was going to the government in the form of
taxes. In contrast Punjab's agricultural sector is

very lightly taxed. There is no income tax and
land revenue forms only a small fraction of the
value of agricultural output (2-3 per cent at
most). Agricultural inputs and outputs are either
lightly taxed or not taxed at all. Thus the only
tax burdens on agriculture are indirect taxes on
urban consumer goods used by the agricultural
sector. On the other hand as we noted in Chapter
III, the state invested heavily in the creation of
infrastructural public goods and the provision of
rural credit which largely benefited the
agricultural sector.
    Thus the product and market contribution of
agriculture in Punjab to the development process
has been at least as much as, if not more than,
that of Japanese agriculture during 1880-1930. In
the case of factor contribution either as taxes or
as savings or in terms of labour force, the Punjab
pattern is almost opposite to that observed in the
process of economic development of Japan.

3 Punjab Growth in an All-India Perspective

Punjab is part of India and there has always been
unrestricted movement of capital and labour between
it and other parts of India. There have been some
periods in the past when official restrictions were
imposed on the movement of food grains between
different parts of India, including Punjab, and
such restrictions were implemented through the
operation of a Zonal System. However, these were
also abolished in the early 1970s. The integration
of the Punjab economy, and its agricultural sector
in particular, with the economies of other regions
of India may therefore be expected to have occurred
to some extent through the functioning of the
market. We have earlier, following Kuznets,
described such integration as an essential part of
the contribution of agriculture to the process of
economic growth. This makes a comparison of
economic growth in Punjab with what has happened
elsewhere in India specially rewarding. In Table
8.2, we show the comparative growth of agricultural
output net of material costs, at constant prices,
in all states for which comparable data are
available. The growth of

Table 8.2 INDEX OF GROWTH OF VALUE ADDED AND CAPITAL STOCK IN THE AGRICULTURAL SECTORS OF INDIAN STATES, 1960-61 TO 1970-71 (1960-61 Prices)

1960-61=100	SDP 1970-71	SDP/LAB 1971-72	CAPITAL 1972	EQUIPMENT 1972	NSA 1972	LABOUR (MALE) 1972
Assam	119.5	N.A.	N.A.	N.A.	N.A.	N.A.
Andhra Pradesh	123.0	105.5	112.1	179.6	104.5	119.0
Kerala	131.2	84.1	148.0	178.3	113.7	155.4
Madya Pradesh	111.1	95.1	115.8	134.8	114.6	121.5
Maharashtra	87.1	72.3	118.5	164.8	95.0	115.7
Gujarat	165.5	119.3	132.9	237.2	99.2	128.4
Karnataka	133.8	115.5	127.3	173.6	101.0	116.6
J. and Kashmir	114.4	N.A.	137.3	180.5	107.5	N.A.
Punjab and Haryana	162.3	122.7	145.1	387.3	106.8	130.9
Rajasthan	101.7	111.7	117.4	176.6	116.4	116.9
West Bengal	117.4	N.A.	N.A.	N.A.	N.A.	N.A.
Uttar Pradesh	117.3	89.6	123.3	173.4	100.8	116.2
Tamil Nadu	101.7	88.2	123.4	286.2	105.9	121.6

Source: Reserve Bank of India Bulletin, April 1978, Dasgupta and Kumari (1983).

agricultural inputs of land, labour and capital
during this period are also reported in this table.
Changes in all variables are reported in index
terms computed, using constant prices whenever
valuation is involved, with the base period of
1960-61. Between 1960-61 and 1971-72, agricultural
output increased most in Punjab (to the tune of 61
per cent), closely followed by Gujarat (where it
increased by 53 per cent). Karnataka, Kerala,
Rajasthan and Andhra Pradesh experienced 'medium'
growth over the period, the increases in
agricultural output being in the range 25 to 30 per
cent. In all other cases with the exception of
Maharashtra agricultural output increased by less
than 20 per cent. In Maharashtra it actually
declined by nearly 16 per cent. The highest
increase in labour productivity in agriculture was
again achieved by Punjab. During 1960-61 to
1971-72 output per male agricultural worker in
Punjab rose by about 23 per cent. Gujarat,
Karnataka and Rajasthan followed in that order with
increases in output per male agricultural worker of
19, 16 and 12 per cent respectively. In Andhra
Pradesh output per male agricultural worker
increased by only 5.5 per cent. In the states of
Madhya Pradesh, Uttar Pradesh, Tamil Nadu, Kerala
and Maharashtra it declined.
     During the same period, net area sown actually
declined in Maharashtra, stayed virtually unchanged
in Uttar Pradesh, Karnataka and Gujarat, and
increased by between 5 and 7 per cent in Tamil
Nadu, Punjab, Jammu and Kashmir and Andhra Pradesh.
The largest increases in net sown area occurred in
Rajasthan, Madhya Pradesh and Kerala.
     Fixed capital in agriculture (exclusive of
land) registered the largest increases in Kerala
and Punjab, 'medium' increases in Jammu and
Kashmir, Gujarat and Karnataka, and small increases
elsewhere. The index of farm machinery and other
agricultural equipment in 1971-72 (with 1960-61 as
100) was 387 in Punjab, 286 in Tamil Nadu and 237
in Gujarat; in other states increases were less
spectacular, though still quite substantial. Taken
together Punjab registered the largest increases in
agricultural output and labour productivity, as
well as use of farm machinery and implements.
Maharashtra on the other hand experienced a decline

Table 8.3 STRUCTURE OF ECONOMIES OF INDIAN STATES, 1960-61 TO 1970-71
(1960-61 Prices)

	1960-61			1970-71		
	A	M+	S	A	M+	S
Andhra Pradesh	58.2	17.8	24.0	52.7	22.5	24.8
Assam	56.5	32.3	11.1	46.2	38.5	15.3
Kerala	55.6	19.2	25.2	50.4	22.1	27.5
Madhya Pradesh	61.7	21.6	16.7	53.0	27.9	19.1
Maharashtra	41.6	32.6	25.6	27.2	42.8	30.0
Gujarat	41.6	31.0	27.4	44.1	30.5	25.4
Karnataka	61.0	24.3	14.7	50.2	31.8	18.0
J. and Kashmir	63.5	12.0	24.5	49.9	19.8	30.3
Punjab and Haryana	59.3	20.4	20.3	56.5	22.3	21.2
Rajasthan	51.9	21.1	27.0	41.1	30.4	28.5
West Bengal	40.5	32.2	27.3	38.3	33.4	28.3
Uttar Pradesh	60.0	16.3	23.7	55.1	20.9	24.0
Tamil Nadu	51.9	21.1	27.0	41.1	30.4	28.5

Source: Reserve Bank of India Bulletin, April 1978.

both in net sown area and in labour productivity in
agriculture and, despite increases in labour and
capital, value-added by its agricultural sector
declined significantly during the period 1960-61 to
1971-72. It is logical to expect that part of
Punjab's food and agricultural products must have
gone through the market channels and the Food
Corporation of India to states like Maharashtra.

Sectoral percentage shares of agriculture,
manufacturing plus transport, and services in the
State Domestic Product of different states of India
for the years 1960-61 and 1970-71 measured at
1960-61 constant prices are reported in Table 8.3.
In 1960-61 Maharashtra, Gujarat and West Bengal
were the only three states in which agriculture's
share of total product was around 40 per cent. In
all other states agriculture accounted for between
50 and 60 per cent of that total product.
Manufacturing plus transport accounted for
one-third of the State Domestic Product in
Maharashtra, Gujarat and West Bengal. In all other
states it was small, accounting for 20 per cent of
the total product or less.

By 1970-71 the picture had changed in Tamil
Nadu and Rajasthan where the share of agriculture
in the State Domestic Product had declined to
around 40 per cent and that of manufacturing and
transport rose to about one-third. A sharp decline
in the share of agriculture in the State Domestic
Product also occured in Karnataka where it fell
from 61 to 50 per cent during 1960-61 to 1970-71.
Correspondingly the share of manufacturing and
transport increased by 7.5 percentage points to
31.8 per cent of the total in 1970-71. All other
states, except Gujarat, registered a decline in the
share of agriculture in the state's total product
by around 3 to 5 percentage points: in Gujarat
agriculture's share increased from 41.6 per cent to
44.1 per cent of the total. In Punjab it declined
by about 2.8 percentage points.

These changes, when seen in the light of
growth of total State Domestic Product and levels
of per capita product reported in Table 8.4,
suggest that it is not necessarily the richer or
faster growing regions of a country which achieve
structural changes in their economy sooner. By
1970-71, the Punjab region was already leading the

rest of India in terms of per capita income. The rate of growth of per capita income has also been higher in this region than elsewhere in India. Yet structural change in the economy has been much less pronounced than in a number of other states, such as Karnataka. The explanation is simple. In a regional context with free movements of labour, capital and products, differential shifts in sectoral shares can occur even in areas with low levels of per capita income or low rates of growth.

4 Political Economy of Development in Punjab

In many less developed countries, the relative importance to be given to industrial and agricultural investment is an important issue of policy. At one stage it was widely believed that industrialisation in such countries is not possible

Table 8.4 GROWTH OF TOTAL AND PER CAPITA SDP IN INDIAN STATES, 1960-61 TO 1974-75 (1960-61 Prices)

	Annual SDP	Growth Rate SDP Per Capita	Per Capita Income 1970-71 Prices
Assam			
Andhra Pradesh	3.51	1.52	578
Kerala	3.57	1.16	566
Madhya Pradesh	2.89	0.36	500
Maharashtra	3.43	0.95	797
Gujarat	2.48	0.38	782
Karnataka	5.20	2.30	522
J. and Kashmir	4.25	1.63	531
Punjab and Haryana	7.0+	5.0+	984
Rajasthan	3.38	0.80	560
West Bengal	2.29	0.06	748
Uttar Pradesh	2.55	0.28	504
Tamil Nadu	1.53	0.05	585

Source: Reserve Bank of India Bulletin, April 1978.

without allocating most if not all investible
resources to the modern industrial sector;
agriculture, on this view, merits attention only at
a later stage. More recently, the reverse sequence
has increasingly found favour with economists and
planners.

In the context of this debate, the nature of
agriculture's contribution to economic development
is of crucial importance. Let us turn, first, to
an aspect of the market contribution of
agriculture, viz. the marketed surplus of food.
Workers to be employed in the non-agricultural
sectors would spend a very large part of their
income on food. This has to come mostly from the
domestic agricultural sector, for possibilities of
importing food in exchange for domestic products
are likely to be limited. It would be difficult to
transfer food through market exchange to the
non-agricultural sector without raising food
prices. This in turn would not only increase
poverty but also, by pushing up money wages in the
organised sector, eat into the investible surplus
of the non-agricultural sector itself, and so slow
down the rate of growth.

One way out would be to rely, in the first
instance, on the factor rather than the market
contribution of agriculture through direct taxes on
agriculturalists' income; reducing their
purchasing power could induce them to sell more
food while indirect taxes on the goods they buy
(agricultural inputs or industrial consumption
goods) would transfer resources out of agriculture
by making these more expensive compared with what
they sell. Such transfer could also be achieved if
savings generated in agriculture were invested in
industry.

Alternatively one could rely primarily on the
product contribution of agriculture, agricultural
surpluses being sought from increased food supply
through higher and more efficient production. For
this purpose substantial investment would be
required in irrigation, and for the production of
modern inputs. Eventually, this would lower food
prices relative to others and make it easier to
transfer food through the market.

In Punjab, it is the last-mentioned strategy
which has been followed. It has met with some

success. Standards of living have appreciably
increased and rural poverty has slowly diminished.
On the other hand, industrial employment has hardly
risen and the structure of production shows only a
little change.

Prospects of further growth in agriculture are
becoming limited as the net area sown approaches
its ceiling. A continued failure of industry to
develop would have serious consequences for
agriculture itself. With off-farm employment
opportunites lagging behind the increase in the
rural population, marginal farmers would be forced
off the land and have to compete with landless
labourers for employment opportunities in
agriculture. As the post-harvesting technology
becomes completely mechanised, employment
opportunities in agriculture will sharply decline.
Conflicts in rural society will become much sharper
than at present and provide a fertile ground for
the further intensification of sectarian and
religious fanaticism.

More attractive futures are possible. Through
their employment in the Indian armed forces,
Punjab's rural youth has had considerable exposure
to new and improved ways of handling economic and
managerial tasks. As a result and especially
because of extensive participation of returned
soldiers in farming activity, Punjab agriculture
has acquired a fairly high level of human capital,
which may not be adequately reflected in
educational statistics. Even in terms of such
statistics, by 1961 Punjab had the highest
proportion of farmers with completed secondary
education among all states of India. One of the
consequences of public investment in infrastructure
and public goods since that time has been a
substantial increase in the incidence of formal
education in both the rural and the urban
workforce.

Further, rural electrification and linking of
all villages with market towns and cities through
the construction of roads and the expansion of
public transport have made the agricultural
workforce in Punjab far more mobile than before and
at the same time given its members a wealth of 'on
the job training' in handling machines and
electrical equipment. This must have generated

skills in efficient handling and maintenance of modern agricultural technology which is in some ways quite similar to industrial technology. Exposure to the army and consequent experience in task-oriented work have ensured that managerial and worker discipline continues to be high. These managerial and worker skills are not so very specific to agriculture and could easily be adapted for efficient use in the processing and manufacturing industries. Human capital in Punjab accumulated through such formal and informal channels seems to be its best resource for embarking on a major industrial take-off; and with the right combination of physical capital invested in selected industries such a take-off could be achieved.

The availability of resources for economic development does not by itself guarantee that they will be effectively utilised for this purpose, for there is not automaticity in the process of either agricultural or industrial development. Since a tradition of modern industrial entrepreneurship has not yet developed in this region, it is inescapable that the state should play an active role in the initial stages of industrial development. In doing so, it should draw on the experience of state intervention for industrial growth elsewhere in India as well as in other countries and take care to avoid restrictive policies of state control, which merely encourage inefficiency and 'rent-seeking'. That the logic of Punjab's development so far now requires accelerated growth of non-agricultural production was made amply clear in our discussion in the preceding chapter. The urgency of the task has been highlighted by the sectarian violence and communal tensions which have built up in the region in mid 1984. The causes of these events are complex and by no means entirely economic in nature; and we do not propose to analyse them here. We should like, however, to make a brief comment.

A rapid process of economic growth and modernisation in a predominantly rural and agrarian society tends to lead to a loosening of ties based on kinship and village, more permissive modes of personal behaviour, a relaxation of customary religious rites and a general weakening of

174

traditional patterns of authority. Frequently such
a process has been observed to produce a revivalist
backlash. Something of the kind appears to have
occurred in parts of Punjab; but the problem seems
to have been aggravated by a purely economic
consideration, viz. the belief by many large and
middle farmers of the region, particularly in that
part of it which corresponds to Punjab state where
the Green Revolution started earlier, that
possibilities of profitable increase in
agricultural production are becoming rapidly
exhausted. In such a situation there are two broad
alternative strategies that could be followed. One
is to retreat and call a halt to the process of
modernisation itself: the other is to take it
forward. The former approach would in our
judgement have disastrous consequences; and it
could lead to a dissipation of the substantial
gains that have been achieved in the process of
development in the past decades. What is required
is a firm, well thought out, and visible effort
towards a more diversified economic structure which
alone could sustain the long-term process of modern
economic development in Punjab. This will not
solve the immediate short-term problem, but that is
the task of political leadership at both the Punjab
and the all-India level.

APPENDICES

Appendix A

STATISTICAL TABLES

Table A.1 CROPPING INTENSITY AND FERTILISER USE IN PUNJAB, 1960-61 TO 1976-77.

	1960-61	1965-66	1971-72	1976-77
Per cent of net irr. area to N.S.A.	42.29	48.80	59.14	63.89
Net irr. area (,000 Ha)	3027	3488	4520	4992
Fertiliser consumption N + P$_2$O$_5$ (,000 Tons)	7.1	49.6	357.4	481.6
Index of fert. N + P$_2$O$_5$	49.34	131.47	282.80	279.43
Index of fert. 1960-61 = 100	100	698.59	5033.80	6783.10
Compound growth rate of fert. consumption per cent per year.	47.52	38.98		6.15

Source: _Statistical Abstracts of India_, various issues.

Table A.2 RURAL POVERTY IN PUNJAB,
1957-58 TO 1977-78

Year	Percentage of rural population in poverty
(1)	(2)
1957-58	28.0
1959-60	24.2
1960-61	18.8
1961-62	22.3
1963-64	29.4
1964-65	26.5
1965-66	26.5
1966-67	29.5
1967-68	33.9
1968-69	24.0
1970-71	23.6
1973-74	23.0
1977-78	17.0

Sources: See Appendix D.

Table A.3 AGRICULTURAL WAGES IN PUNJAB, 1950-51 TO 1974-75

	1950-51*		1956-57*		1964-65*		1974-75*	
Average Daily Earnings (Rs.) of Men and Children Belonging to Agricultural Labour Households	Men	Children	Men	Children	Men	Children	Men	Children
(a) In Agricultural Occupation	1.84	0.98	1.98	0.69	2.13	1.04	6.72	3.45
(b) In non-Agricultural Occupation	1.82	1.85	1.38	0.66	2.07	1.26	5.86	3.09
a:b	1.01	0.53	1.43	1.05	1.03	0.83	1.11	1.15
In Different Agricultural Occupation: Ploughing	1.84		2.08	1.50	2.05	0.80	5.81	3.43
Harvesting	2.68		2.47	0.45	2.66	1.15	6.97	4.50

* Relates to casual agricultural labourers only.
Source: Chaudhri (1979) and Statistical Abstracts of Punjab and Haryana, various issues.

Table A.4 SEX RATIO IN PUNJAB 1901 to 1981

State	1901	1911	1921	1931	1941	1951	1961	1971	1981
Haryana State	867	835	844	844	869	871	868	867	870
Punjab State	832	780	799	815	836	844	854	865	879
All India	972	964	955	950	945	946	941	930	933

Source: Census of India 1971 Series I - India Part II-A (i) and Census of India 1981 Series I.

Appendix B

INDEX NUMBERS OF AGRICULTURAL OUTPUT

Series A : These were derived from the data on State Domestic Product by industry of origin published in the Reserve Bank of India Bulletin, April 1978 and May 1979, and prepared by the National Income Studies Division of the Department of Statistics, Reserve Bank of India, on the basis of official estimates of State Domestic Product. Agricultural output is measured net of cost of materials and also of depreciation. These represent, in our judgement, the best available estimates of value added in agriculture at the state level. Unfortunately, they are not available for years other than those shown in Table 2.1. From our point of view, it would have been preferable to use data net of material costs but gross of depreciation since the capital stock data (discussed in Appendix C) are measured gross of depreciation. Unfortunately, these were not available.

Series B: Up to 1965-66, these are the index numbers of agricultural output, as published in the Statistical Abstracts of Punjab, and adjusted to the base year 1960-61. The index numbers from 1966-67 onwards are weighted averages of the Punjab and Haryana indices, published in the Statistical Abstracts of these states. The weights used, viz. 62.5 (Punjab) and 37.5 (Haryana), represent their relative contributions to their combined total agricultural output in the sense of Series A, averaged over the years 1969-70, 1970-71 and 1971-72.

Series C & D:These represent the aggregate value, relative to the base year, of ten principal crops of the region, viz. rice, wheat, barley, jowar, bajra, maize, gram, sugarcane, American cotton and Desi cotton, which account on average for over 90 per cent of total crop output. Both series evaluate crops at Punjab from harvest prices. Series C uses for each crop its harvest price averaged over the years 1969-70, 1970-71 and 1971-72, and D the average for 1976-77, 1977-78 and 1978-79. Data on quantities and prices were taken from various issues of the Statistical Abstracts of Haryana and Punjab.

We have used 1960-61 as the base year for these as also for some other indices used in this study. This seemed an appropriate choice not only because 1960-61 was a 'normal' year for crop production conditions as well as in other respects, but also because it makes it easier to compare our results with those of other studies on Indian agriculture, which frequently use the same base year.

Appendix C

ESTIMATES OF AGRICULTURAL CAPITAL

Implements and Machinery

The estimates for these were derived by the
so-called 'direct' method. The amounts, in physical
units, of each item were obtained from the
Quinquinnial Livestock Censuses, held in 1951, 1956,
1961, 1966 and 1972. They were evaluated at
constant 1970-71 prices collected by the Central
Statistical Organisation (CSO), Government of India,
representing average market prices for Punjab –
except for a few items of 'other modern equipment'
for which CSO prices were not available. These were
evaluated at prices obtained from a sample survey of
200 rural households in Ludhiana (Punjab) and Karnal
(Haryana) in 1971 by the Agro-Economic Research
Centre, University of Delhi; again, the prices are
market prices. Our estimates thus imply a
replacement value concept of capital and no
allowance is made for depreciation of any part of
the stock. All items for which the quantity data
are available for all the Livestock Census years
have been included in our estimate.

Items included under 'other modern equipment'
were recorded for the first time in the Livestock
Census of 1966; but for the earlier years it could
safely be assumed that their use was negligible.
The following items are included in this category:
improved threshers, maize threshers,
seed-cum-fertiliser drills (tractor and power
operated), improved seed drills, disc harrows and
cultivators, improved harrows and cultivators, power
tillers, sprayers and dusters. In a few cases,
there are changes in definition between the 1966 and
1972 Censuses. To ensure comparability, figures
were adjusted in the light of Table 2.
[(footnote),Indian Livestock Census, Vol.1, 1972,
p159.]

Livestock

The same method of estimation was used as in the
case of implements and machinery. The numbers of

different categories of working and milch animals
given by the Livestock Censuses were evaluated at
1970-71 prices as given by CSO. The working animals
included are: (i) male cattle above three years of
age used for work only; (ii) male buffaloes above
three years of age used for work only; (iii) camels
above four years of age. Milch animals include cows
and she-buffaloes.

There are a number of points to be made about
the definition of working animals that we have used.
First, a small number of cows and she-buffaloes are
also used as working animals but we have omitted
them because the data are not recorded in a
consistent way between different Livestock Censuses.

Secondly, the Livestock Censuses from 1961
onwards provide a classification of male working
animals into those used for breeding only, those
used for work only, and those used both for work and
breeding. The earlier Censuses do not, however,
provide such a breakdown. In order to maintain
comparability we have included only the animals that
are used exclusively for work. Since the number of
animals used both for breeding and for agricultural
work is relatively small, neglecting them is
unlikely to make much difference to our results.

Lastly, we have not distinguished between urban
and rural animals. Our figures include both. Using
these for measuring inputs for agricultural
production clearly requires some justification. The
justification essentially is that 'urban' areas, as
defined by the Livestock Census, include a large
number of small towns and large villages in which
livestock is used for the purpose of farm production
(Shukla 1965, p.68). It is only in the big
metropolitan areas that work animals such as cattle
or buffaloes are used predominantly for the purpose
of carrying loads; but the numbers involved are
relatively small. These considerations led Shukla
to conclude that the exclusion of livestock in the
urban areas would lead to a larger error than their
inclusion. We agree with her judgement.

Estimates of Agricultural Capital

Farm Buildings

We considered two alternative ways of evaluating
rural buildings (of which farm buildings are a
component). One would be to use the same method as
that for implements and for livestock. This was
rejected because of certain difficulties. Although
the total number of census houses (rural) is given
in each decennial Population Census, the definition
of a census house has not remained uniform. Thus,
the 1951 definition of a 'Census House' as a
'dwelling with a separate main entrance' was
considerably broadened in 1961 (see Census of India,
1961, Punjab, Part IV-A, Report on Housing and
Establishments, p.7; and Census of India, 1971,
Punjab Part IV, Housing Report and Tables, Appendix
III). Using the Census figures would, therefore,
considerably overstate the increase in the constant
price value of housing during 1951-61. Data on
prices are also far from satisfactory. For these
reasons we decided to use the inventory method as
described below.

For 1961-62, the All-India Rural Debt and
Investment Survey reported in the
Reserve Bank of India Bulletin, June 1965, gives
state-specific estimates of the aggregate value of
tangible assets in rural areas as well as the
percentage share in this total of various items and,
in particular, that of 'House Property' consisting
of (i) residential buildings and (ii) other
structures and building sites (Statement 3, p.33;
Table II, p.4). The value of rural buildings in
Punjab was calculated directly from these. The
estimate is based on market prices: for each item
of tangible assets, the market value as on 30 June
1962 given by the respondent was used.

For 1971, we used the All India Debt and
Investment Survey, 1971-72 Statistical Tables,
Haryana, Himachal Pradesh, Punjab and Delhi, the
Reserve Bank of India. Table 3 Haryana and Table 3
Punjab of this publication give the average value of
buildings per rural household in Haryana and Punjab
respectively. The estimated aggregate numbers of
rural households in the two states are also given in
the table facing page 4. The total value of rural

buuildings in Punjab and Haryana together are
computed from these. The underlying concepts are
essentially the same as in the 1961 survey; and the
figures refer to the values of assets as on 30 June
1971.

For 1950-51, the lack of a comparable study
makes it infeasible to derive an estimate on similar
lines. The best that we could do was to adjust the
1961 figure downwards. The Census of India, Punjab,
1961, Part IV-A, Report on Housing and
Establishments, tells us that the number of
'dwellings' in rural Punjab increased by 20.9 per
cent during 1951-61 (p.16), that unlike census
houses they are inter-censally comparable (pp.7,16);
and that they accounted for nearly two-thirds of all
rural census houses in 1961 (p.12). It would not be
unreasonable to assume that the percentage increase
of all rural census houses was the same as that of
dwellings. By adjusting the estimate already
derived for 1960-61 accordingly, we obtained the
value of rural buildings in 1950-51 at 1960-61
prices. The estimates for 1950-51 as well as for
1960-61 were then converted into their equivalent in
1970-71 prices, using the Index Numbers of
Investment Cost for Construction (Rural) (1960-61 =
100), prepared by the CSO and given in Uma Datta Ray
Chaudhri, 'Industrial Breakdown of Capital Stock in
India', Journal of Income and Wealth, April 1977,
Table 4, p.154. Figures for 1955-56 and for 1965-66
were derived by (logarithmic) interpolation.

We have taken a third of the total value of
rural buildings to represent their 'agricultural
capital' component. This figure was derived as
follows. The Census of India, Punjab, 1961, Part
IV-A, observes that in rural Punjab, every fourth
census house is used 'chiefly for tethering
livestock and dry fodder' (p.12). Since such
structures are less costly than residential
buildings, their share of the total value would
certainly be less than 25 per cent. This is
supported by the data of Table II, p.4, of the All
India Rural Debt and Investment Survey according to
which for rural households (all-India), out of the
total value of 'Residential Buildings plus other
structures and building sites' the latter had a

186

share of just over 18 per cent. We can use this figure for Punjab as well. However, in addition, part of the residential building itself may be used for housing cattle, storing fodder, seed and implements, and for feeding family or hired labour. Assuming this part to be a fifth leads to our 'one-third' rule. Some earlier studies such as Mukherjee and Shastry (1974), Shukla (op.cit.) and the Reserve Bank of India (1974) have also made the same assumption, but without spelling out a justification. Some scholars on the other hand have criticised such an approach on the grounds that 'there is no very firm basis on which to separate the value of the farm residence from that of other buildings' (Tostlebe (1957) p.5). He even maintains that 'any division between productive and consumptive uses of the farm residence would be altogether arbitrary and open to questions' (ibid, p.5), since the farm residence not only provides an abode convenient to the fields and barns for the farmer and his family but also serves to feed and to house hired help (a point to which we have already referred).

This criticism appears to us to be exaggerated and Tostlebe's own alternative of including the full value of housing property in the estimate of agricultural capital grossly inflates the capital stock in agriculture.

A more legitimate reason for concern is the evidence from recent survey data, the Cost of Cultivation Studies for example, that investment in residential housing proper has been growing faster than in such items as farm houses, cattlesheds and 'golas'. If this trend reflecting perhaps the life-style of richer farmers prevails, using the one-third rule for projection may lead to the future value of farm buildings being over estimated. For the period up to 1971-72, there is little reason to worry on this account.

Appendix D

A NOTE ON THE POVERTY INDEX

In the context of measuring poverty in India, the
minimum ('poverty line') has been specified in terms
of a monthly per capita expenditure, which
represents the cost of a bundle of basic goods. It
varies between the Indian states and also between
rural and urban areas. The poverty line used in
deriving estimates of poverty reported in Table 6.1
for years other than 1977-78 is Rs 15.9 per capita
per month in 1960-61 rural Punjab prices, which is
equivalent to an average all-India rural poverty
line of Rs 15: the higher figure for Punjab
reflects regional price differences. This figure
was derived in the following way.
Paterardham(1963), following earlier work by Ay
Kryod, had recommended a balanced diet providing
minimum nutrition required for an average adult male
in moderate activity as consisting of 15 oz.
cereals, 3 oz. pulses, 4 oz. milk, 1.5 oz. sugar
and gur, 1.25 oz. edible oils, 1 oz. ground nut
and 6 oz. vegetables per day. This gives 2100
calories and 5.5 grams of protein per day. Bardhan
(1974), using NSS rural retail prices, derived the
cost of this diet, with minor adjustment due to
non-availability of appropriate price data for
ground nuts and vegetables and also to allow for
lower prices for non-monetised consumption. The
cost of nutrition required per adult was converted
into a corresponding figure per person, using an
adult equivalent ratio. To this was added a cost
component for non-food items.
    No well-defined norms being available for such
items, their cost was derived from actual

expenditure data of a 'reference' population, viz.
the lowest five deciles of the rural population, and
was computed simply by multiplying the cost of the
'minimum' diet already derived by the ratio of
non-food to food expenditure for this population,
calculated from NSS data. The total of the food and
non-food cost so derived represents the poverty
line. To make (rural) poverty lines for different
years comparable, the Consumer's Price Index for
Agricultural Labour (General Prices) was used for
conversion. For years up to 1973-74 percentages of
the rural population below the poverty line were
computed from NSS data in this way by Ahluwalia
(1978) and are reported in Table 6.1.

The figure for 1977-78 is a weighted average of
the Planning Commission's estimates of the rural
poverty ratio in Punjab and Haryana states, using
their rural population as weights. The estimates
were derived using esentially the same method as
described above but based on a poverty line
corresponding to a slightly higher calorie norm.
Some difficulties are mentioned below.

Setting a cut-off point for poverty is
necessarily arbitrary in some degree. However, it
is worth noting that, using a different approach,
Rajaraman (1975) arrived at a poverty line of Rs
16.36 per capita per month in 1960-61 rural Punjab
prices, which is only slightly higher than ours.
She too treated its food and non-food components
differently. The former was defined as the minimum
cost of a diet, with prices given, that satisfied
constraints imposed by minimum nutritional
requirements, taste and local availability and was
derived by linear programming. To this she added a
small allowance for items of food without nutritive
value, such as tea and spices, and the cost of
non-food items (20.2 per cent of the total), derived
from NSS data on actual consumption in 1960-61 of
the pooret 30 per cent of the rural population.
Using this poverty line, Rajaraman derived estimates
of rural poverty in Punjab for 1960-61 and 1970-71,
viz. 18.40 per cent and 23.28 per cent which again
are close to the estimates we have used.

Secondly, NSS data on consumption from which
the number of people below the poverty line is
calculated have their own difficulties, some of
which bearing on the inequality measure were

A Note on the Poverty Index

considered earlier. The fact that estimates of aggregate consumption by NSS tend to differ from those published in National Income Statistics has also often been commented on. That such differences are due to systematic bias in NSS data has recently been argued by Tyagi (1982), who believes that NSS estimates of aggregate food consumption for the early 1960s were overestimates; and that the degree of overestimation diminished over time as a result of a greater vigilance in recording food consumption and general improvement in the quality of investigation in later years. This, it is argued, may have led to an upward bias in the time-trend of poverty as computed. However, errors in estimating the consumption of the poor may not necessarily have followed the same time-path as those affecting aggregate consumption; and the National Income estimate of consumption, which is computed as a residual, is not believed to be particularly reliable either.

More worrying than these is the index number problem. The CPIAL, as remarked earlier, is based on the consumption mix of agricultural labourers in 1956-57 and its failure to allow for substitution, especially between coarse cereals and wheat, may have led both to an overestimation of the true poverty line and hence of the incidence of poverty in later years (Tyagi, op.cit.), and to an exaggeration of year-to-year fluctuations in poverty.

As against this it has been stated that the CPIAL agrees closely with fractile-specific or expenditure-class specific indices reflecting the recent consumption mix of the rural poor (Bardhan, 1974; Datta,1978) On closer scrutiny, such agreement is found only over certain sub-periods. Murthy and Murthy (1977) have computed a comprehensive series of price indices for each decile class of the per capita household expenditure distribution in rural India for the years 1952-53 to 1973-74. We took the average of their price indices for the lowest three deciles and compared this with the CPIAL over 1955-56 to 1973-74. For the earlier part of the period, they show similar magnitudes of price increase but later, say from 1967-68 to 1973-74, the rise in CPIAL is very much greater, and it also shows year-to-year fluctuations of greater

A Note on the Poverty Index

amplitude.

On balance, we believe that the estimates we have used somewhat underestimate the decline in the poverty ratio especially since the late 1960s.

More detailed evidence on the decline of poverty in Punjab is given in Bhalla and Chaddha (1982) and Mundle (1982). Problems of defining the poverty line in India, espcially those of data, are discussed in a number of papers in Srinivasan and Bardhan (1974). That definitions based on an average nutritional requirement, such as the one used here, overestimate the incidence of poverty by failing to allow for interpersonal variability of nutritional needs is forcefully argued by Sukhatme (1977, 1981). Different concepts of poverty and the problems of measurement they involve are lucidly surveyed in Kakwani (1980).

GLOSSARY

BAJRA	Indian corn, millet
DESI	Native, not foreign
GAUNA	The ceremony of taking a bride to her husband's house for the first time
GHAIRMUMKIN	Impossible
GOTRA	Family, clan, lineage
GUR	Raw sugar
JATI	Community, caste
JOWAR	Indian millet
KACHA ARATHIYA	Auctioneer
KUTCHA	Clay-built, not strong
MISL	Territory or Principality under Sikh chiefs
MUKLAWA	Same as GAUNA
PUCCA ARATHIYA	Buying agent
RAGI	Finger-millet
RYOTWARI SYSTEM	Where the government is regarded as absolute proprietor of the land, and the ryot or cultivator paid a fixed revenue, according to acreage, directly to the government
VARNAS	Caste, four principal classes described in Manu's code, viz. Brahmins, Kshatriyas, Vaisyas and Sudras.

BIBLIOGRAPHY

ADELMAN, I., and C.T. MORRIS (1973) Economic
  Growth and Social Equity in Developing Countries,
  Stanford University Press.
AGARWAL, B., (1977) Mechanization in Farm Operational
  Choices and Their Implications: A Study Based on
  Punjab. Ph.D. Dissertation, University of Delhi.
AGARWAL, B., (1980) 'Tractorization, Productivity
  and Employment - A Reassessment', Journal of
  Development Studies, vol. 16.
AGRICULTURAL ECONOMICS RESEARCH CENTRE, (1970)
  Primary Education in Rural India Participation
  and Wastage, Tata McGraw-Hill Pub. Co., New
  Delhi.
AHLUWALIA, M.S., (1974) 'Income Inequality: Some
  Dimensions of the Problem' in Chenery, Hollis
  et al.(Ed.), Redistribution With Growth, Oxford
  University Press, London.
AHLUWALIA, M.S., (1976) 'Inequality, Poverty and
  Development', Journal of Development Economics,
  December.
AHLUWALIA, M.S., (1978) 'Rural Poverty and
  Agricultural Performance in India', Journal of
  Development Studies, April.
ALLCHIN, B. and R. ALLCHIN, (1968) The Birth of
  Indian Civilization, Penguin Books, Harmondsworth.
ANDERSON, J.R. and J.L. DILLON, (1971) 'Allocative
  Efficiency, Traditional Agriculture and Risk',
  American Journal of Agricultural Economics,
  February.
BARDHAN, P.K., (1970) 'On the Minimum Level of
  Living and the Rural Poor', Indian Economic
  Review, April.
BARDHAN, P.K., (1970) 'The Green Revolution and
  Agricultural Labourers', Economic and Political
  Weekly, Special Number, July.
BARDHAN, P.K., (1972) 'On the incidence of poverty

in rural India', Economic and Political Weekly,
Annual Number.
BARDHAN, P.K., (1973) 'Size, Productivity, and
Returns to Scale: An Analysis of Farm-Level
Data in Indian Agriculture', Journal of
Political Economy, November-December.
BARDHAN, P.K., (1979) 'On the Incidence of Poverty
In Rural India in the Sixties' in Srinivasan
and Bardhan (eds.), Poverty and Income Distri-
bution in India, Indian Statistical Institute,
Calcutta.
BARDHAN, P.K. and T.N. SRINIVAS, (1974) 'Crop-
Sharing Tenancy in Agriculture - Rejoinder',
American Economic Review, Vol. 64.
BASHAM, A.L., (1953) The Wonder That Was India,
Grove Press, New York.
BERRY, I. and W.R. CLINE, (1979) Agrarian Structure
and Productivity in Developing Countries, Johns
Hopkins University Press, Baltimore.
BHADURI, A., (1973) 'Agricultural Backwardness
under Semi-Feudalism, Economic Journal, March.
BHAGWATI, J.N. and S. CHAKRAVARTY, (1969) 'Contri-
butions to Indian Economic Analysis: a Survey',
American Economic Review vol. 59(4), Pt.2,
Supplement.
BHAGWATI, J.N. and T.H. SRINIVAS, (1974) 'Reanalyzing
Harris-Todaro Model. Policy Rankings in Case of
Sector-Specific Sticky Wages', American Economic
Review, vol. 64.
BHALLA, G.S., (1982) The Green Revolution in the
Punjab (India), Jawaharlal Nehru University,
New Delhi, mimeographed.
BHALLA, G.S. and G.K. CHADDHA, (1983) Green
Revolution and the Small Peasant: A Study of
Income Distribution among Punjab cultivators,
Concept Publishing Company, New Delhi.
BHARDWAJ, K., (1974) Production Conditions in
Indian Agriculture, University of Cambridge
Department of Applied Economics, Occasional
Paper 33, Cambridge University Press, London.
BILLINGS, M.H. and A. SINGH, (1970) 'Mechanisation
and Rural Employment: With Some Implications
for Rural Income Distribution', Economic and
Political Weekly.
BILLINGS M.H. and A. SINGH, (1971) 'The Effect of
Technology on Farm Employment in India',
Development Digest, vol. 9, January.
BIRNBERG, T.B. and S.A. RESNICK, (1975) Colonial
Development, Yale University Press, New Haven.

BOSERUP, E., (1965) The Conditions of Agricultural
    Growth, Allen and Unwin, London.
BOSERUP, E., (1981) Population and Technology,
    Basil Blackwell, London.
BLYN, G., (1966) Agricultural Trends in India 1891-
    1947: Output, Availability and Productivity,
    University of Pennsylvania, Philadelphia.
BRAHME, S., G.R. SAINI, D.D. NARULA and B. PRAKASH,
    (1979) 'Integrated Rural-Development Program
    with Special Reference to Chandrapur', Indian
    Journal of Social Work, vol. 39.
BROWN, D.D., (1967) The Intensive Agricultural
    Districts Programme and Agricultural Development
    in Punjab, India, Economic Development Report
    No. 79, presented at the D.A.S. Conference,
    Sorrento.
CHAUDHRI, D.P., (1968) Education and Agricultural
    Productivity in India, Ph.D. Dissertation,
    University of Delhi.
CHAUDHRI, D.P., (1974) 'New Technologies and
    Distribution in Agriculture', in D. Lehmann (ed.)
    Agrarian Reform and Agrarian Reformism, Faber and
    Faber, London.
CHAUDHRI, D.P., (1979) Education, Innovations and
    Agricultural Development, Croom Helm, London.
CHAUDHRI, D.P., (1979) 'Factor Proportions and
    Technical Change in Punjab Agriculture'. A paper
    presented at the Australian Society of Agricultural
    Economics Annual Conference, Canberra.
CHENERY, H.B. and M. SYRQUIN, (1975) Patterns of
    Development, 1950-1970, Oxford University Press,
    London.
CLARK, G., (1969) World Prehistory: A New Outline,
    Cambridge University Press, Cambridge.
CLINE, W.R., (1977) 'Policy Instruments for Rural
    Income Redistribution' in Frank, Charles R., Jnr.,
    and Webb, Richard C., Income Distribution and
    Growth in the Less Developed Countries, The
    Brookings Institution, Washington, D.C.
COMMERCE (INDIA), (1972) Annual Number: Regional
    Profile of Indian Agriculture, vol. 125.
CUMMINGS, R.W., (1967) Pricing Efficiency in the
    Indian Wheat Market, Impex, India.
DANDEKAR, V.M. and N. RATH, (1971) Poverty in India,
    Indian School of Political Economy, Poona.
DARLING, M.L., (1925) The Punjab Peasant in Pros-
    perity and Debt, Oxford University Press, London.
DASGUPTA, A.K., (1981) 'Demographic Trends in India;
    An Economic Interpretation', Working Paper.

Bibliography

Institute of Economic Growth, New Delhi.
DASGUPTA, A.K., (1982) 'Agricultural Growth Rates
    in the Punjab, 1906-42', Indian Economic and
    Social History Review, July-December.
DASGUPTA, A.K., (1984) Growth and Composition of
    Agricultural Capital Stock in Indian States,
    1961-77, IEG Occasional Paper, New Series No. 9,
    Hindustan Publishing Corporation, Delhi.
DASGUPTA, A.K. and N.S. SIDDHARTHAN, (1984)
    'Industrial Distribution of Indian Exports and
    Joint Ventures Abroad', Institute of Economic
    Growth, New Delhi.
DATTA, B., (1978) 'On the Measurement of Poverty
    in Rural India', Indian Economic Review, April.
DAVIES, K., (1951) Population of India and Pakistan,
    Princeton.
DAVIES, S., (1979) The Diffusion Process of Inno-
    vations, Cambridge University Press, Cambridge.
DAY, R.H., and I. SINGH, (1977) Economic Development
    as an Adaptive Process, The Green Revolution in
    Indian Punjab. Cambridge University Press,
    Cambridge.
DEATON, A., and J. MUELLBAUER, (1980) Economics and
    Consumer Behaviour, Cambridge University Press,
    Cambridge.
DENISON, E.J., (1962) The Sources of Economic
    Growth in the United States and the Alternatives
    Before US. Committee for Economic Development,
    New York.
ETIENNE, G., (1976) Agricultural Growth and Rural
    Development in Indian and Pakistan Punjabs,
    Institute of Development Studies, Geneva.
    December. Mimeographed.
FIELDS, G.S., (1981) Poverty Inequality and Develop-
    ment, Cambridge University Press, Cambridge.
FRANK, C., Jr. and R.C. WEBB (eds.) (1977) Income
    Distribution and Growth in the Less-Developed
    Countries, Brookings Institution, Washington.
FISK, E.K., (1962) 'Planning in a Primitive Economy:
    Special Problems of Papua New Guinea', Economic
    Record, vol. 38, pp. 462-78.
FISK, E.K., (1964) 'Planning in a Primitive Economy:
    From Pure Subsistence to Production of a Market
    Surplus', Economic Record, vol. 40, pp. 156-74.
GIRI, I., (1962) 'Land Records and Land Use
    Statistics', Agricultural Situation in India,
    December.
GIRI, I., (1966) 'Changes in Land Use Pattern in
    India', Indian Journal of Agricultural Economics,

Bibliography

July-September.
GOVERNMENT OF INDIA, (1952) Report of the Grow More
    Food Enquiry Committee, New Delhi.
GRIFFIN, K., (1981) Land Concentration and Rural
    Poverty, Macmillan, Second Edition, London.
GUPTA, D.B., (1982) (ed.) Development Planning and
    Policy: Essays in Honour of Professor V.K.R.V.
    Rao, Wiley Eastern Limited, New Delhi.
GUPTA, D.B., (1982) Rural Industry in India: The
    Experience of the Punjab Region, Occasional
    Paper No. 7, Hindustan Publishing Company, Delhi.
HABIT, I., (1972) Proceedings of the Punjab History
    Conference, 1971, Patiala.
HALAN, Y.G. and P.B. DESAI,(1982) 'Rural Industry
    in India - The Experience of the Punjab Region',
    Occasional Paper, New Series, No. 7, HPC, Delhi.
HANUMANTHA RAO, C.H., (1975) Technological Change
    and Distribution of Gains in Indian Agriculture,
    Macmillan, Delhi.
HANUMANTHA RAO, C.H.,(1979) Farm Mechanisation
    Forming Part of Agricultural Development of
    India: Policy and Problems, Ed. by C.H. Shah,
    Agricultural Development of India: Policy and
    Problems, New Delhi, Orient Longman, 1979, xii,
    688p.
HARCOURT, G.C., (1969) 'Some Cambridge Controversies
    in the Theory of Capital', Journal of Economic
    Literature, vol. 7.
HAYAMI, Y. et al.,(1975) A Century of Agricultural
    Growth in Japan: Its Relevance to Asian Develop-
    ment, University of Tokyo, Japan.
HAYAMI, Y. and V. RUTTAN, (1971) Agriculture and
    Economic Development in International Perspective.
    Johns Hopkins University Press, Baltimore.
HEADY, E.O., (1951) Economics of Agricultural
    Production and Resource Use.
HEADY, E.O. and S.T. SOUNKA, (1973) 'Farm size,
    rural-community income, and consumer welfare',
    American Journal of Agricultural Economics,
    vol. 56.
HELLER, W.W., (1976) Economy: Old Myths and New
    Realities, W.W. Narton, New York.
HICKS, J.R., (1957) Value and Capital, Oxford
    University Press, London.
HICKS, J.R., (1959) Essays in World Economics,
    Oxford University Press, London.
HICKS, J.R., (1981) Nobel Lecture 'The Mainspring
    of Economic Growth', Swedish Journal of Economics,
    December.

HILTON, R., (1973) Bound Men Made Free, Methuen,
London.
HIRSCHMAN, A.O., (1958) The Strategy of Economic
Development, Yale University Press, New Haven.
ISHIKAWA, S., (1981) Essays on Technology, Employment
and Institutions in Economic Development,
Comparative Asian Experience. Economic Research
Series No. 19, Institute of Economic Research,
Hitotsubashi University, Tokyo.
JOHNSTON, B.F., (1966) 'Agriculture and Economic
Development in Japan', Stanford Food Research
Institute Bulletin, Stanford.
JONES, E.L., (1974) Agriculture and Industrial
Revolution, Basil Blackwell, Oxford.
JUSSUWALA, M.F., (1969) Economics of Development,
Oxford and IBH Publishing Co., Calcutta.
KAHLON, A.S. et al., (1973) Impact of Mechanisation
on Punjab Agriculture With Special Reference to
Tractorisation, Punjab Agricultural University,
Ludhiana.
KAHLON, A.S. and G. SINGH, (1967) 'Economics of
Power Wheat Thresher', Indian Farm Mechanization,
vol. 13.
KAHLON, A.S. and S.S. MIGLANI, (1972) Studies in
the Economics of Farm Management, Ferozepur
District (Punjab), Combined Report, 1967-68 to
1969-70, Punjab Agricultural University, Ludhiana,
Mimeographed.
KAHLON, A.S. and D.S. TYAGI (1983) Agricultural
Price Policy in India, Allied Publishers, New
Delhi.
KAKWANI, N., (1980) Income Inequality and Poverty:
Methods of Estimation and Policy Applications,
Oxford University Press, London.
KESSINGER, T.G., (1974) Vilayatpur, 1848-1968:
Social and Economic Change in a North Indian
Village, University of California Press, Berkeley.
KOSAMBI, D.D., (1965) The Culture and Civilization
of Ancient India in Historical Outline, Routledge
and Kegan Paul, London.
KRISHNA, R., (1964) 'The Growth of Aggregate
Agricultural Output in the Punjab', Indian
Economic Journal, July-September.
KRISHNA, R., (1975) 'Measurement of the Direct and
Indirect Employment Effects of Agricultural
Growth with Technical Change', in Earl O. Heady
and Larry R. Whiting. Externalities in the
Transformation of Agriculture: Distribution of
Benefits and Costs of Development, The Iowa State

University Press, Ames.

KRISHNAMURTY, K., (1966) 'Economic Development and Population Growth in Low Income Countries: An Empirical Study for India', Economic Development and Cultural Change, October.

KUMAR, D., (ed.), (1982) The Cambridge Economic History of India, vol. 2, c.1757-c.1970, Cambridge University Press, Cambridge.

KUZNETS, S., (1955) 'Economic Growth and Income Inequality', American Economic Review, March.

KUZNETS, S., (1966) Modern Economic Growth: Rate, Structure and Spread, Yale University Press, New Haven.

LEAF, M.J., (1983) 'The Green Revolution and Cultural Change in a Punjab Village, 1965-78', Economic Development and Cultural Change, January.

LEWIS, G.B., (1978) Quantitative Methods in Economics, University Press, Sydney.

LELE, U.J., (1971) Foodgrains Marketing in India - Private Performance and Public Policy, Cornel University Press, Cornel.

LEWIS, W.A., (1976) 'Development and Distribution', in Alec Cairncross and Mohinder Puri (Ed.), Employment, Income Distribution and Development Strategy, Macmillan, London.

LEWIS, W.A., (1978) Growth and Fluctuations 1870-1913, Allen and Unwin, London, Evolution of the International Economic Order, Kalyani Publishers, New Delhi. (Elist Janeway lectures on historical economics in honour of Joseph Schumpeter, 1977)

LEWIS, W.A., (1978) Evolution of the International Economic Order, Princeton University Press, Princeton.

LIPTON, M., (1977) Why Poor People Stay Poor: A Study of Urban Bias in World Development, Maurice Temple Smith Ltd., London.

LYDALL, H.F. and L.C. THUROW, (1977) 'Generating Inequality', Economic Journal, vol. 87.

MCLEOD, W.H., (1976) The Evolution of the Sikh Community, Clarendon Press, Oxford.

MADDISON, A., (1970) Economic Progress and Policy in Developing Countries, George Allen and Unwin Ltd., London.

MARX, K., (1887) Capital vol. 1, Foreign Languages Publishing House, Moscow, 1954.

MAZUMDAR, D., (1963) 'On the Economics of Relative Efficiency of Small Farmers', Economic Weekly, Special Number.

MAZUMDAR, D., (1965) 'Size of Farm and Productivity:

A Problem of Indian Peasant Agriculture',
Economica, vol. 32.
MELLOR, J.M., (1968) 'The Evolution of Rural
Development Policy', in J.M. Mellor et al.
Developing Rural India, Lalvani Publishing House,
India.
MICHELL, A.A., (1967) The Indus River, A Study of
the Effects of Partition', Yale University Press,
New Haven and London.
MINHAS, B.S., (1970) 'Rural Poverty, Land Distribution
and Development', Indian Economic Review, April.
MITRA, A., (1977) Terms of Trade and Class Relations,
Frank Cass, London.
MORAWETZ, D., (1974) 'Employment Implications of
Industrialisation in Developing Countries: A Survey',
Economic Journal, September.
MORGENSTERN, O., (1963) On the Accuracy of Economic
Observation, second edition, Princeton University
Press, New Jersey.
MUELLEBAUER, J., (1974) 'Prices and Inequality -
United Kingdom Experience', Economic Journal,
vol. 84.
MUKHERJEE, M. and N.S.R. SHASTRY, (1974) 'An Estimate
of the Tangible Wealth of India' in R. Goldsmith
and C. Saunders (eds.): The Measurement of National
Income, Income and Wealth Series VIII, Bowes &
Bowes, London, 1959.
MUNDLE, S., (1982) Land, Labour and Level of Living
in Rural Punjab, I.L.O., Geneva. Mimeographed.
MURTY, G.V.S.N. and K.N. MURTY, (1977) 'On Differ-
ential Effects of Price Movement', Indian
Economic Review, October.
MUSHRAFF, J., (1980) Growth in Crop Output: Pakistan
and Indian Punjabs Since 1947, Ph.D. Dissertation,
Australian National University.
NADIRI, M.I., (1970) 'Some Approaches to the Theory
and Measurement of Total Factor Productivity: A
Survey', Journal of Economic Literature,
December.
NAIK, J.P., (1974) Policy and Performance in Indian
Education, New Delhi, Orient Longman, 1975,
112 p. (K.G. Saiyidain Memorial Lectures, 1974)
NURKSE, R. (1952) Problems of Capital Formation in
Underdeveloped Countries, Oxford University Press,
London.
NURKSE, R., (1961) 'Further Comments on Professor
Rosenstein-Rodan's Paper', in H.S. Ellis, and
H.L. Wallich, (eds.), Economic Development in
Latin America, Macmillan, London pp. 640-3.

OHKAWA, K., (1965) 'Agriculture and the Turning
    Point in Economic Growth', The Developing
    Economies, December.
OHKAWA, K. and H. ROSOVSKY, (1968) 'Postwar Japanese
    Growth in Historical Perspective' in L. Klein
    and K. Ohkawa (eds.) Economic Growth: The
    Japanese Experience since the Meiji Era, Richard
    D. Irwin Inc., Homewood, Ill.
OHKAWA, K. and SHINOHARA (eds.), (1979) 'Patterns
    of Japanese Economic Development - A Quantitative
    Appraisal', London, Yale University Press, 1979,
    xiv, 411 p. (Economic Growth Center Publications)
PATRICK, H. and H. ROSOVSKY (eds.), (1976) Asia's
    New Giant, Brookings Institution, Washington,
    D.C.
PATTERSON, K.D. and K. SCHOTT, (1979) The Management
    of Capital, Theory and Practice, Macmillan,
    London.
PATVARDHAN, V.S. (1964) 'A Note on Data Relating
    to Production and Marketing in Rural Credit
    Follow-up Surveys of the Reserve Bank of India',
    Arthvijnana 8, 4, Dec. pp. 260-72.
PAUKERT, E., (1973) 'Income Distribution at
    Different Levels of Development: a Survey of
    Evidence', International Labour Review, August-
    September.
PAUL, S. and A.K. DASGUPTA, (1983) 'Inheritance
    and the Distribution of Wealth in Punjab (India)',
    Working Paper, Delhi School of Economics,
    University of Delhi.
PRABHA, C., (1969) 'District-wise Rates of Growth
    of Agricultural Output in Pre-Partition and
    Post-Partition Punjab', Indian Economic and
    Social History Review, December.
PYATT, G. and WALTER GALENSON, (1964) The Quality
    of Labour and Economic Development in Certain
    Countries, ILO, Geneva.
PYATT, G.F.S., (1964) Priority Patterns and the
    Demand for Household Durable Goods, Cambridge
    University Press, Cambridge.
RAO, V.K.R.V., (1974) 'New Challenge Before the
    Indian Agriculture', Journal of the Indian
    Society of Agricultural Statistics, 26(1),
    June 1974; pp. 33-6
RAO, C.H.H. and K. SUBBARAO,(1976) 'Marketing of
    Rice in India: An Analysis of the Impact of
    Producer's Prices on Small Farmers', Indian
    Journal of Agricultural Economics, 31(2),
    April-June 1976.

Bibliography

RAJARAMAN, I., (1975) 'Poverty, Inequality and
    Economic Growth: Rural Punjab, 1960-61 to
    1970-71', The Journal of Development Studies,
    July.
RANDHAWA, M.S., (1974) Green Revolution: A Case
    Study of Punjab, Vikas, Delhi.
RAYCHAUDHURI, T. and I. HABIB (eds.), (1982) The
    Cambridge Economic History of India Volume I,
    c.1200-c.1750, Cambridge University Press,
    Cambridge.
RAYCHOUDHRI, U.D., (1977) 'Industrial Breakdown of
    Capital Stock in India', Journal of Income and
    Wealth, April.
REYNOLDS, L.G., (1977) Image and Reality in Economic
    Development, Yale University Press, New Haven.
ROY, P., (1981) 'Transition in Agriculture:
    Empirical Indicators and Results (Evidence from
    Punjab, India)', Journal of Peasant Studies,
    vol. 8.
RUBEN, W., (1974) 'Outline of the Structure of
    Ancient Indian Society', in R.S. Sharma (ed.),
    Indian Society: Historical Probings, People's
    Publishing House, Delhi.
RUDRA, A., (1974) 'Minimum Level of Living, A
    Statistical Examination' in Srinivasan and Bardhan.
RUDRA, A., (1982) Indian Agricultural Economics.
    Myths and Realities, Allied Publishers, New Delhi.
SABERWAL, S., (1976) Mobile Men, Limits to Social
    Change in Urban Punjab, Vikas, New Delhi.
SAINI, G.R., (1969) 'Resource-Use Efficiency in
    Agriculture', Indian Journal of Agricultural
    Economics, April-June.
SAINI, G.R., (1979) Farm Size, Resource-Use
    Efficiency and Income Distribution, Allied
    Publishers Private Ltd., New Delhi.
SAITH, A., (1981) 'Production, Prices and Profit
    in Rural India', Journal of Development Studies,
    January.
SARKAR, GAUTAM K., (1983) Commodities and the Third
    World, Delhi, Oxford University Press, 1983,
    viii, 182 p.
SARKAR, H. and E.O. HEADY, (1979) 'Production
    Response of High-Yielding Variety of Rice in
    Some Less Developed Countries - Estimation and
    Comparison', Indian Journal of Agricultural
    Economics, vol. 34.
SAWADA, S., (1969) 'Technological Change in Japanese
    Agriculture: A Long-Run Analysis', in K. Ohkawa,
    B.F. Johnston and H. Kaneda (eds.), Agriculture

and Economic Growth: Japan's Experience,
University of Tokyo Press, Tokyo.

SCHOLLIERS, P., and C. VANDENBROKE, (1982) 'The
Transition from Tradition to Modern Patterns of
Demand in Belgium' in Henri Bandet and Henk van
der Meulen (eds.), Consumer Behaviour and
Economic Growth in the Modern Economy. Croom
Helm, London and Canberra.

SCHULTZ, T.W., (1953) The Economic Organization
of Agriculture, McGraw Hill, New York.

SCHULTZ, T.W., (1980) Nobel Lecture, 'On Economics
of being poor', Journal of Political Economy,
March-April.

SEN, A.K., (1962) 'An Aspect of Indian Agriculture',
Economic Weekly, Annual Number.

SEN, A.K., (1964) 'Size of Holdings and Productivity',
Economic Weekly, February.

SEN, A.K., (1964) 'Working Capital in India', in
P. Rosenstein Roden (ed.), Pricing and Fiscal
Policies, M.I.T. Press, Cambridge, Mass.

SEN, A.K., (1973) On Economic Inequality, Clarendon
Press, Oxford.

SEN, A., (1975) Employment, Technology and
Development, Oxford University Press, London.

SEN, A.K. and R. AMJAD, (1977) 'Limitations of a
Technological Interpretation of Agricultural
Performance - A Comparison of East Punjab (India)
and West Punjab (Pakistan), 1947-72', South Asia
Papers, Nov-Dec 1977, South Asian Institute,
University of Punjab, Lahore.

SHARMA, R.K., (1982) Draught Power Planning in
Indian Agriculture: A Case Study of Haryana,
Ph.D. Dissertation, University of Delhi.

SHUKLA, T., (1965) Capital Formation in Indian
Agriculture, Vora & Co., Bombay.

SIMON, J.L., (ed.), (1980) The Relationship Between
Population and Economic Growth in LDCS, Jai Press,
Greenwich.

SINGH, I.J. and S.S. JOHL, (1966), Field Crop
Technology in the Punjab, Social Systems Research
Institute, University of Winconsin, Madison.

SINGH, H.K.M., (1979) 'Population Pressure and
Labour Absorbability in Agriculture and Related
Activities. Analysis and Suggestions based on
Field Studies as Conducted in Punjab', Economic
and Political Economy, vol. 14.

SINGH, H.K.M. and A.S. OBERAI, (1980) 'Findings of
a Case Study in the Indian Punjab', International
Labour Review, vol. 119.

SINHA, J.N., (1982) '1981 Census economic data', Economic and Political Weekly, February 1982.

SMITH, A., (1776) The Wealth of Nations, Edited by E. Cannan (1937), Modern Library, New York.

SOLOW, R.M., (1957) 'Technical Change and the Aggregate Production Function', Review of Economics and Statistics, vol. 39.

SOLTOW, L., (1968) 'Long-Run Changes in British Income Inequality', Economic History Review, vol. 31, No. 1.

SOLTOW, L., (1969) Six Papers on the Size Distribution of Wealth and Income, National Bureau of Economic Research, New York.

SRINIVASAN, T.N. and P.K. BARDHAN (eds.), (1974) Poverty and Income Distribution in India, Statistical Publishing Society, Calcutta.

SRIVASTAVA, U.K., V. NAGADEVARA and E.O. HEADY, (1973) 'Resource Productivity, Returns to Scale and Farm Size in Indian Agriculture: Some Recent Evidence', Australian Journal of Agricultural Economics, April.

SUKHATME, P.V. and B.V. SUKHATME, (1970) Sampling Theory of Surveys with Applications, Revised Edition, Asia Publishers, Bombay.

SUKHATME, P.V., (1977) 'Malnutrition and Poverty', Lal Bahadur Shastri Memorial Lecture, Indian Council of Agricultural Research, New Delhi, Mimeographed.

SUKHATME, P.V., (1981) 'On Measurement of Poverty', Economic and Political Weekly, August.

SUMIYA, M. and K. TAIRA, (1979) An Outline of Japanese Economic History 1603-1940: Major Works and Research Findings, University of Tokyo Press, Tokyo.

TEITELBAUM, M.S., (1974) 'Population and Development: Is a Consensus Possible', Foreign Affairs, vol. 52.

THAPAR, R., (1975) The Past and Prejudice, National Trust, New Delhi.

THAVRAJ, M.J.K., (1963) 'Pattern of Public Investment in India', Indian Economic and Social Historical Review, July-September.

TOSTLEBE, A.S., (1957) Capital in Agriculture, Princeton University Press, Princeton.

TSURU, S., (1958) Essays on Japanese Economy, Kinokuniya Book Store, Tokyo.

TYAGI, D.S., (1982) 'How Valid are the Estimates of Trends in Rural Poverty?', Economic and Political Weekly, 26 June.

VERMA, B.N., (1980) Infrastructural Gap and Emerging

Regional Disparity in Economic Performance in
Indian Economy, Nineteenth Indian Econometric
Conference (30 December 1980 to 1 January 1981),
Poona.

UNICEF, (1981) An Analysis of the Situation of
Children in India, UNICEF Regional Office, New
Delhi.

WHEELER, M., (1966) Civilizations of the Indus
Valley and Beyond, Thames and Hudson, London.

WYOH, J.B., et al., (1966) 'Delayed Marriage and
Prospects for Fewer Births in Punjab Villages',
Demography, vol. 3, No. 1.

YASUO, K., (1942) Economics of Japanese Agriculture,
Jichosha, Tokyo.

YUOTOPOULOS, P.A. and L.J. LAU, (1971) 'A Test
for Relative Efficiency and Application to Indian
Agriculture', American Economic Review, 61, 11,
pp. 94-109, March.

YOUNGSON, A.J., (1967) Overhead Capital, Edinburgh
University Press, Edinburgh.

OFFICIAL REPORTS AND PERIODICALS CITED

Government of India

Agricultural Handbook, ICAR (1961)
Agricultural Labour Enquiry Report 1964-65
Census of India, 1951 to 1981, various reports
District Census Reports for Punjab and Haryana
Districts
Farm Management Studies Reports for Punjab, various
issues
Handbook of Housing Statistics, CBO (1980)
Health Statistics of India, various issues
Indian Livestock Census Reports 1956 to 1972
National Sample Survey Reports, various rounds
Reserve Bank of India Bulletin, various issues
Royal Commission on Agriculture 1928, various
volumes of the report
Rural Labour Enquiry Report 1974-75
Sample Registration Bulletin, various issues
Statistical Abstracts of India, various annual
issues
Vital Statistics of India, Civil Registration
Statistics, various issues

Governments of Punjab and Haryana

Punjab Alienation of Land Act 1900
Punjab Banking Enquiry Commission Report 1929

Official Reports and Periodicals cited

Statistical Abstracts of Haryana, various annual issues

Statistical Abstracts of Punjab, various annual issues

Year Book of Agricultural Statistics, Haryana, 1971-72

Year Book of Agricultural Statistics, Punjab, 1971-72

INDEX

For Product Safety Concerns and Information please contact our EU
representative GPSR@taylorandfrancis.com Taylor & Francis Verlag GmbH,
Kaufingerstraße 24, 80331 München, Germany

Printed and bound by CPI Group (UK) Ltd, Croydon, CR0 4YY

01/05/2025

01858464-0002